Collins

need to know?

Weather
Watching

Patrick Hook

Collins

First published in 2006 by Collins
an imprint of
HarperCollins Publishers
77–85 Fulham Palace Road
London W6 8JB

www.collins.co.uk

A catalogue record for this book is available from
the British Library

Created by **Focus Publishing**
Project editor: Guy Croton
Designer: David Etherington
Series design: Mark Thomson

ISBN-13 978 0 00 722300 8
ISBN-10 0 00 722300 5

Colour reproduction by Colourscan, Singapore
Printed and bound by Printing Express Ltd,
Hong Kong

Contents

1 Weather basics

The fortunes of humanity have been linked to the vagaries of the weather since time immemorial. Knowing whether it is going to be hot or cold, wet or dry, is just as important today as it was for our ancestors. Extreme conditions – such as severe storms or prolonged drought – can lead to starvation, sickness and death, and still do in many parts of the world. As evidence of climate change becomes ever more compelling, never before has good knowledge of climate and weather conditions been quite so significant to the peoples of the world.

Early forecasting

Trying to work out what the weather is going to do has been a matter of great importance ever since mankind first evolved. We don't know much about what our earliest ancestors did in this respect, but from recorded times onwards weather forecasting has played a central part in many societies.

The origins of forecasting

The Bible has numerous references to the subject – the old adage that 'a red sky at night is a shepherd's delight and a red sky in the morning is a shepherd's warning' is just one example. In the Hebrew *Book of Job*, written around the 5th century BC, the Lord asked Job a series of questions about the weather. These included such matters as where rain, lightning, ice, frost and the east winds came from. The ancient Greeks were very interested in the

The saying 'Red sky at night – shepherd's delight, red sky in the morning – shepherd's warning' is thousands of years old.

weather, and had many different prediction methods, varying from examining the entrails of animals through to direct observation. Aristotle, for instance, wrote a philosophical treatise in the 4th century BC on the subject, in which he presented his ideas on all manner of naturally occurring phenomena ranging from snow to rainbows. This was entitled *Meteorologic*, and was the first serious study of the atmosphere ever conducted. The title derived from the Greek word *meteorol*, which refers to something that happens in the sky. The name is still used today in the form of the English word 'meteorology', which means 'the study of the atmosphere'. The Romans were also keen on foretelling the weather, although after Aristotle, there were few significant advances in the science until the time of the Renaissance in the 16th century, when Galileo Galilei invented the thermometer. About fifty years later Evangelista Torricelli – one of Galileo's proteges – developed the mercury barometer.

Weather lore

There are hundreds of old sayings that concern the weather, and since the written word has only become available to the public in the last few hundred years, they were composed in such a way as to make them easier to remember. Some are in the form of lengthy poems, whereas others are short proverbs. For instance, there is a 16th-century rhyme about St Swithin's Day, which is celebrated every year on 15 July; St Swithin was a 9th-century bishop of Winchester, England:

St Swithin's day, if thou dost rain,
For forty days it will remain;
St Swithin's day, if thou be fair,
For forty days 'twill rain na mair.

The proverb simply says that if it is raining on 15 July, then it will rain for forty days, but if not, then it will be dry for forty days.

Seaweed has been used for centuries as an aid to predicting the weather.

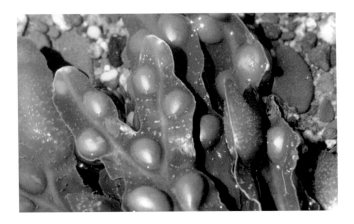

Lore concerning the animal and plant world

A large proportion of weather lore is based on the animal and plant world. The following is a selection of favourite old sayings:

▶ *Seagull, seagull sit on the sand. It's never good weather when you're on the land.*
▶ *When the trees begin their dance. Of rain there is a great big chance.*
▶ *Moss dry, sunny sky, moss wet, rain we will get.*
▶ *Bees do not swarm before a storm.*
▶ *Cows lying down, good indication of rain.*
▶ *Flowers smell best just before the rain.*

Clouds and the sky

Many old sayings concern the clouds and state of the sky:

▶ *The higher the clouds the better the weather.*
▶ *Mackerel sky and mares' tails make lofty ships carry low sails.*
▶ *When clouds appear like rocks and towers, the earth's refreshed with frequent showers.*

Nocturnal observations

Some of the old sayings concern observations made at night:

▶ *A change in the moon brings on a change in the weather.*
▶ *If all stars are out at night it will be a nice day tomorrow.*

- *When halo rings the moon or sun, rain's approaching on the run.*
- *If the moon holds water it will be dry. If water from it can leak rain is nigh.*
- *If you see a ring of clouds around the moon, it's going to rain within a day.*

The wind

Winds have inspired many a longstanding weather proverb:
- *When the wind is from the south the rain's in its mouth.*
- *When the wind is blowing in the North, no fisherman should set forth,*
  When the wind is blowing in the East, 'tis not fit for man nor beast,
  When the wind is blowing in the South, it brings the food over the fish's mouth.
- *Wind from the east fish bite least, wind from the west fish bite best.*
- *The wind in the west suits everyone best.*

Rain and fog

There are even sayings that are specific to rain and fog:
- *If it's foggy in the morning then it'll be a sunny day.*
- *Rain before seven, clear by eleven.*
- *When dew is on the grass, no rain will come to pass.*
- *Rainbow at night, sailors delight; rainbow in the morning, sailors warning.*
- *If there is a heavy dew, it will not rain.*

Perhaps one should not pay too much attention to weather proverbs, though – while they may be accurate in some places at some times, they do not work in all places at all times. A good example is the saying that *'The wind in the west suits everyone best'*. While this may be true along eastern coasts where such winds come from over the land, on western coasts they blow straight off the sea, and therefore bring cold, wet weather with them.

Pine cones are another natural weather forecasting aid that have been widely used for many years.

The birth of modern forecasting

It was not until Robert Hooke of London – known variously as 'The Inspirational Father of Modern Science' as well as 'The English Father of Meteorology' – focussed his attentions on the subject that weather prediction became a science.

The innovations of Robert Hooke

The results of some of Hooke's activities in this area were published in his magnificent work *Micrographia*. Widely recognized as one of the greatest scientific books ever published, it is best known for the large number of incredibly detailed drawings he made of the minute creatures he saw through a microscope. What is not so well known, is that the book also presented many meteorological instruments that he either invented or improved, along with detailed instructions on how to use them. These included the wheel barometer, rain gauges, a humidity meter or 'hygrometer', wind speed recorders and devices which recorded continuous weather measurements. One of his many legacies is that he was the first to suggest that the freezing point of water should be used as the zero point on the thermometer.

An early Signal Service weather map on record in NOAA Library – dated 1 September 1872. Careful compilation and study of these maps led to ever more scientific forecasting, which was first pioneered by Robert Hooke in the 17th century.

A US Navy bi-plane of the early 20th century with a meteorograph on its starboard wing strut, taking meteorological measurements for pressure, temperature and humidity.

As Robert Hooke continued to improve his ways of measuring and recording the weather, he realized that forecasts might be possible if enough readings were taken and then tabulated accurately. From his detailed studies and endless observations, he convinced himself that hurricanes, storms, mists and fogs were all the result of changes in air pressure. He therefore laid down the foundations of the science that we know today as meteorology.

Weather-forecasting today

These days weather-forecasting is much more precise – we do not need to check whether seaweed is drying out, or if pine cones are opening up. Instead, we can rely on a wide range of modern techniques. The various parameters that affect the weather – factors such as temperature, pressure and rainfall – are all measured and recorded very precisely by a whole array of scientific instruments. The data are then analysed, and a long series of weather statistics are generated using the very latest

US Army Air Force meteorologists prepare to launch a hydrogen-filled balloon.

computational methods. When enough calculations have been made, it is possible to make predictions as to what the weather is likely to do. This is an incredibly complex procedure, with a huge number of permutations to take into account. It should be remembered therefore, that a prediction is just that – we are still a long way from being able to say exactly what the weather will do for anything more than a few hours ahead.

Climate and the weather

The terms 'weather' and 'climate' have different meanings. The word 'weather' describes the day-to-day variations of the atmospheric conditions of a given location. The word 'climate' – which comes from the Greek *klima* – is a summation of the overall effects of the weather over a long period. The planet's climate is therefore ultimately determined by the behaviour of the weather systems of the various geographic regions around the world. These in turn reflect the effects of the sun's radiation on their respective atmospheres and surfaces. The amount of the sun's heat that reaches a particular place is governed by a number of factors – these include the latitude (that is, the distance from the equator), the altitude, the atmospheric conditions and the local geography.

Infrared imagery of upper Chesapeake Bay, as obtained by the Landsat satellite.

Changing climates

These days, it seems that the term 'climate change' is on everyone's lips; however, the world has a long history of such changes. Ice ages have occurred frequently – these are long-term periods during which there is significant global cooling. This causes the world's ice sheets to extend dramatically, and large glaciers to form across much of what are currently temperate regions. The first ice age occurred around 2.7 to 2.3 billion years ago, and the most recent one ended only about 10,000 years ago.

The atmosphere has a major influence on the amount of heat that the Earth receives from the sun. If it contains large amounts of greenhouse gases such as carbon dioxide and methane, then heat from the sun gets trapped and global warming occurs. If, on the other hand, there are large amounts of particulates in the air, then a larger proportion of the sun's heat gets reflected back into outer space. This can happen when large volcanic eruptions take place.

The way that the Earth orbits the sun is also a potential candidate for changes in global temperatures, as both changes in the orbital eccentricity and axial inclination have direct effects.

The positions of the continents change over time, and many scientists think that when there are sufficient land masses near the poles, this phenomenon can also trigger an ice age.

must know

Icy for ages
Over the last few million years there have been many ice ages. For a considerable time they happened every 40,000 years or so, but since then the interglacial periods have increased to 100,000 years.

A three-dimensional cloud-top image of Hurricane Diana created from data captured by the TIROS-N satellite in 1984.

Climate zones

The long-term weather patterns of a given area are used to categorize its prevailing climate. These patterns include the seasonal variations, as well as any extremes experienced such as droughts, hurricanes or blizzards.

Categorizing climate zones

The amount of precipitation and average air temperatures are the most important factors used to determine a region's climate type. The fauna and flora that are found in a given area are dictated to a large extent by its climate – together the three components form what is known as a 'biome'. There are many different biomes around the world. Some are too cold for all but the hardiest of organisms to survive, whereas others are tropical and teem with myriad species of flora and fauna. In between there are various other areas, each with their own characteristic climates – these are divided up by what is known as the Köppen Climate Classification System. This was devised in 1900 by the Russian-German climatologist Wladimir Köppen, and divides the world's climates up into five basic types, each of which is designated by a capital letter (see opposite page).

Other ways of categorizing climate zones

The Köppen System is not the only way of categorizing climates – there are several others besides. For instance, it is possible to divide the world's climates into tropical, sub-tropical, temperate, cold and polar zones.

The amount of annual precipitation is also a major determinant of an area's climate – there are eight basic categories. These are: equatorial; tropical; semi-arid tropical; arid; dry Mediterranean; Mediterranean; temperate and polar. While the temperature levels and amounts of precipitation are

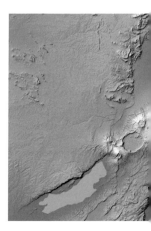

This image showing the geography of the vast savannah plains of the Serengeti was created from data obtained by the Shuttle Radar Topography Mission (SRTM) aboard the Space Shuttle Endeavour.

convenient criteria by which to divide up the world's climate zones, neither method is an accurate guide to the types of flora and fauna that can be found within them. This method of classification is much more subjective, but is still useful for identifying the nature of each zone. It includes tropical rainforests, savannahs, deserts, steppe, chaparral or Mediterranean climates, dry grasslands, deciduous forests, moist continental climates, polar and Arctic, tundra, taigas, boreal forest and alpine or highland climates.

Different climate zones of the world

The diversity of the world's climate zones is extraordinary, with each type displaying unique characteristics and defining qualities which clearly separate it from any other.

Tropical rainforests

Tropical rainforests can be found in the equatorial belt, between the tropic of Cancer (23°27' north) and the tropic of Capricorn (23°27' south). This area covers about 40 per cent of the Earth's surface, and includes much of south-east Asia where the forests of Indonesia, Papua New Guinea, and the East Indies are amongst the most exotic in the world. Parts of Australia, Africa, China, India and Central and South America also have huge areas of tropical rainforest, with the Amazon Basin being one of the best known examples. These areas all have moist tropical climates with high temperatures (around 27–35°C/81–95°F) and large amounts of rainfall throughout the year – often exceeding 250 cm (100 inches) annually. This creates a very humid atmosphere, with values of between 80 and 90 per cent being the norm. As a result of the high temperatures and humidity, cumulus clouds often start to form by midday.

Tropical rainforest features large amounts of rainfall combined with high temperatures – an environment in which many plants and animals thrive.

Most savannahs are found between the latitudes of 25oS and 25oN, and exhibit marked seasonal changes, with a very wet rainy season followed by a very hot dry season.

Savannahs

The word 'savannah' comes from the 16th century-Spanish term *zavanna*, which means 'treeless plain'. Although savannahs are, as the translation suggests, open tropical grasslands that are generally treeless, some have a scattering of shrubs and small trees, whereas others support light forestation. They form in areas that have enough rainfall to support desert grasses, but too little to permit the growth of forests. This is usually between 100mm and 400mm (4 and 16 inches), most of which falls within a few months.

Dry grasslands

Dry grasslands are similar to savannahs in many ways, but they are located nearer the poles, and as a result have much greater annual temperature variations. While their summers vary from being warm to hot, the winters are always cold, and rainfall is low at around 81 cm (32 inches) a year. These climates are only found in the centres of large continental masses where wet air from the oceans is held back by mountain ranges. This unusual combination of conditions means that the largest areas of dry grasslands are only found in western North America.

Deserts, which support a very specialized range of wildlife, can be found in many parts of the world. This spectacular example is in South America.

Deserts

Deserts – areas with very low annual precipitation levels – can be formed through a combination of factors. The oldest deserts are due to a blend of geological features and climatic conditions; some of the more recent ones, however, are the direct result of human activity. The most arid kinds of deserts cover around 12 per cent of the world's surface – these places have annual rainfall figures of only around 2.5mm (0.1 inches), and are found between the latitudes of 15°–25° N and S. This includes much of the south-western United States and northern Mexico, as well as parts of Argentina, north Africa, south Africa and the central part of Australia. There are other deserts, however, which have higher annual precipitation levels – up to 250mm (10 inches) or so. These are also very dry as a result of high average temperatures and evaporation rates that exceed the amount of rainfall. The lack of moisture in the air means that the humidity levels are too low for clouds to form. This exacerbates the problem, as the ground receives most of the heat from the sun's rays – daytime temperatures can reach 55°C (131°F) in the shade. At night, the lack of cloud cover results in most of the heat accumulated during the day being radiated away, and temperatures fall to around freezing. This wide temperature range and lack of water produces a harsh environment for plants in which only a few species can survive, and consequently there is a paucity of soil cover. Since there is so little to feed on, few large animals are found in the desert. Those that are able to eke out an existence tend to be highly specialized for life in arid surroundings – examples include beetles, scorpions, snakes and small rodents.

Steppe

The steppe is a semi-arid climate region that is typified by vast grasslands. These are only found in the interiors of the large land masses of the North American and Eurasian continents, where wet air from the oceans is blocked by mountain ranges. The amount of precipitation is therefore low, and varies

The steppe is a harsh place to live due to the extremes of temperature and low rainfall. This image shows the extensive grasslands of Mongolia as seen by the Moderate-resolution Imaging Spectroradiometer (MODIS), on NASA's Terra spacecraft.

between 10cm (4in) in the driest areas, to 50cm (20in) in the wetter zones. If the levels were to fall below this, the steppe would not be able to support grasses, and would soon become arid desert. If they were wetter, taller growing species of grasses would take over, and the steppe would be quickly replaced by a tall-grass prairie. The summers are generally warm or hot, but during the winter, cold air from the poles flows down over the steppe. This is unable to escape from the region as its path is blocked by the same mountains that prevented rain from reaching it in the summer. As a result the temperatures fall to well below freezing, and remain low for many months at a time. The main areas of steppe are found in two zones between the latitudes of 35° and 55°N. One of these is in western North America – across the Great Basin, the Columbia Plateau, and the Great Plains. The other is in the Eurasian continental interior, and can be found in vast swathes that are distributed from eastern Europe to the Gobi Desert and northern China.

Mediterranean or chaparral climate
The Mediterranean or chaparral climate is very hot and dry in the summer – reaching a peak of around 38°C (100°F), and mild and moist in the winter, with temperatures rarely falling below freezing. The annual precipitation is low, with levels of between 25cm (10in) and 42cm (17in). Most of this falls in the winter months, and the long periods of hot, dry weather cause the region's vegetation to become desiccated. As a result, only plants that can cope with prolonged droughts are able to survive, and wildfires are common. Indeed, many plant species have evolved to depend on the heat from such fires to open up their seed pods. Mediterranean climates are typified by biomes where cork oaks, pine, olive, and eucalyptus trees flourish, and can be found between the latitudes of 30°–50° North and South. The main areas are in central and southern California, the coasts that

border the Mediterranean Sea, coastal Western and South Australia, the Chilean coast, and the Cape region of South Africa.

Moist continental climate

The moist continental climate is found in the zone where cold air from the polar regions meets hot air from the tropics. Where the two fronts collide, rain clouds form in profusion, and precipitation is usually high throughout the year, at around 81 cm (32in). Winters tend to be cold and wet as polar winds predominate, but in the summer the tropical winds take over and temperatures can soar. These climates are usually found between the latitudes of

This composite image of the Iberian peninsula was constructed from data captured by the Multi-angle Imaging SpectroRadiometer (MISR) on NASA's Terra satellite.

30°–55° North and South, although in Europe they lie between 45°–60° North. They are located in eastern parts of the United States, as well as southern parts of Canada, northern China, Korea, Japan and central and eastern Europe.

Taiga

The taiga – or boreal forest – is a continental climate that has long, harsh winters, and short, cool summers. It is found between the latitudes of 50°–70° North and South, and so borders the polar regions. As a result, the main influences are the cold air masses which flow down from the world's ice caps, giving this region the highest temperature variations of any climate zone. In the winter, these cold fronts cause temperatures to fall as low as -25°C (-14°F). In the summer, the climate is much milder, with highs of around 16°C (60°F). There is little rain, however, and annual precipitation is only around 31 cm (12in). On the American continent, taigas can be found across central and western Alaska, from the Yukon Territory to Labrador in Canada. In Eurasia, they are spread across northern Europe and Siberia, as far as the Pacific Ocean.

OPPOSITE: **The moist continental climate is typified by very lush broadleaf woodlands where high rainfall results in numerous small streams and rivers.**

This image produced by the Earth Observations, Image Analysis Laboratory, at the Johnson Space Centre shows Lake Teletskoye in Siberia, which is surrounded by ancient taiga forests.

The tundra is a harsh, but beautiful place. Sadly, even these remote regions are being damaged by pollution and changing climates.

Tundra

Tundra climates are only found along the coasts of the Arctic region between the latitudes of 60°–75° North. This covers much of the northernmost parts of the North American continent, and includes the Hudson Bay region, the coasts of Greenland, and the areas of northern Siberia that border the Arctic Ocean. Tundra zones exhibit a marked seasonality, with long, harsh winters, however, even though they are further north than the taigas, they do not get as cold in winter. This is due to the stabilizing influence of the oceans – lows of around -22°C (-10°F) are experienced. There is no true summer as such – instead there is a short mild season, which has highs of only 6°C (41°F) or so. Precipitation is low, with only around 20cm (8in) of rain falling each year.

OPPOSITE: **This image of the snow-capped mountain chain that forms California's eastern border with Nevada was taken by NASA's Terra satellite on 31 October 2004.**

Alpine or highland climates

Alpine or highland climates, which are found at high altitudes on mountains, are highly changeable, and get colder the higher they are. They are distributed across the world, and tend to have similar

This beautiful image of South Georgia Island, which lies 1,300 kilometres (807 miles) to the east of the Falkland Islands was produced by the Image Science & Analysis Group, Johnson Space Centre.

seasons to those of the lower altitude regions around them. They do, however, generally experience much wider temperature variations, with winter lows of around -18°C (-2°F) and summer highs of 10°C (50°F). Although they are often shrouded in cloud, these regions do not receive a lot of precipitation, with annual levels in the order of 23cm (9in) or so. When snow falls, it usually accumulates through the winter months, and then begins to melt in the spring. For many of the biomes found along the waterways below, this yearly release of water is vital to their existence. Examples of Alpine climates include the Rocky Mountain Range in North America, the Altai mountain range in Central Asia, the Andes in South America, the Alps in Europe, and the Himalayas in Tibet.

The polar zones

At the centre of the Arctic is the North Pole, which is the northernmost point of the axis around which the Earth rotates. The Pole itself is not on land, though – it is in the middle of the Arctic Ocean. At these latitudes, the sun never sets at midsummer, and never rises above the horizon at mid-winter. During the winter it is a place of extremely harsh conditions where only the hardiest of creatures can survive. Vast areas of the sea freeze solid, forming great swathes of pack ice as the temperatures fall to -70°C (-94°F) and high winds scour the region. As winter gives way to summer, most of the sea ice melts and the environment warms up considerably. Antarctica, which covers more than 13 million sq km (8 million sq miles), is the area south of the Antarctic Circle, centred on the South Pole. It is the coldest place on Earth, with the lowest temperature ever recorded at Vostok in 1983, of -89.2°C (-128.6°F). In the middle of winter so much of the sea around it freezes into pack ice that the continent effectively doubles in size. A perhaps unexpected consequence of the extreme cold is that in many parts of Antarctica there is very little snow. This is because the low temperatures cause nearly all of the water vapour in the air to freeze soon after it comes off the sea. Some of the inner regions of the continent are so dry that they are in effect cold deserts.

want to know more?

Take it to the next level...

▶ **Basic observation** 136
▶ **Cloud classification** 81
▶ **Long-range forecasting** 139
▶ **Global warming** 160

Other sources...

▶ **Look at the sky as the sun rises or sets - see what colour it is, then see what the weather does.**
▶ **Find a pine cone and put it somewhere outside where you can watch it as the weather changes.**

Weblinks...

For more information about the weather covered in this chapter, visit the following websites:

▶ **www.nssl.noaa.gov**
▶ **www.nasa.gov/centers /johnson/home/index. html**
▶ **www.bbc.co.uk/ weather/weatherwise/ living/effects/lore. shtml**

The JOHN N. COBB 400m (¼ mile) from Lamplugh Glacier, Glacier Bay. Date: September 1992.

2 Wind

The atmosphere is the layer of gases that surround the Earth – they are held in place by the planet's gravitational pull. From top to bottom, the atmosphere is about 560 kilometres (350 miles) deep, and is vital to life on our planet. For a start, it provides air for us to breathe, but it also protects the planet's surface from harmful ultraviolet radiation, recycles water from vapour into rain, and renders many toxic chemicals harmless. On top of all this, the atmosphere insulates the Earth from the freezing vacuum of outer space.

Winds and the atmosphere

The air we breathe today has taken about 4.5 billion years to evolve – it was not until about 570 million years ago that there was enough oxygen available for respiring animals to survive in the oceans. About 170 million years later, this had increased to the point where it could support air-breathing land animals.

This photograph of the crescent moon was taken by crewmembers from the Space Station, about 360 km (225 miles) above the Earth.

The make-up of the atmosphere

Our present atmosphere is largely made up of gases that have been released by various planetary processes, such as volcanic activity. It is possible, however, that comets from outer space may also have contributed to it. Most of the oxygen originally came from primitive plants called cyanobacteria, which are also known as blue-green algae. The largest component of dry air is nitrogen at around 78 per cent, with oxygen making up about 21 per cent. Most of the remaining 1 per cent is comprised of argon, although there are many other gases in the atmosphere. These include the various greenhouse gases, such as carbon dioxide (0.01–0.1 per cent) and ozone (0–0.01 per cent), as well as trace amounts of hydrogen, methane, carbon monoxide, helium, neon, krypton and xenon. It must be remembered that these figures are for dry air – depending on the atmosphere's humidity level, up to 7 per cent can be made up of water vapour.

Atmospheric pressure

The atmosphere's pressure is not constant – it varies with many factors, especially height, decreasing as altitude increases. This is because gravity pulls the molecules that make up the atmosphere towards the Earth's surface, and so at low levels there is a considerable column of air pressing down from above. As one travels up this column the weight decreases, which causes a corresponding lowering of the air pressure. Atmospheric pressure is therefore the force exerted by the weight of the air, and is measured by a barometer – hence the term 'barometric pressure'.

The layers of the atmosphere
Although the periphery of the Earth's atmosphere is more than 547 kilometres (340 miles) from the ground, most of it – 80 per cent or so – is within the first 16 kilometres (10 miles). Where the atmosphere blends into outer space, there is so little air that there is no pressure. However, as one moves towards the Earth's surface, it gradually increases. At an altitude of 10,000 feet, there is a pressure of around 10 pounds per square inch (psi), and at sea level, the pressure climbs to about 14.7 psi. The atmosphere is generally divided up into five distinct zones – different people, however, use different conventions. The divisions used here are:

1) Troposphere 2) Stratosphere 3) Mesosphere 4) Thermosphere 5) Exosphere.

The troposphere
The troposphere is the layer closest to the surface of the Earth, and is known as the 'lower atmosphere'. It holds around half of the world's atmospheric gases, and is where all the activities we call 'weather' take

This photograph taken from space on 4 March 1996 shows a haze of industrial air pollution, dust and smoke from China spreading out over the Pacific Ocean.

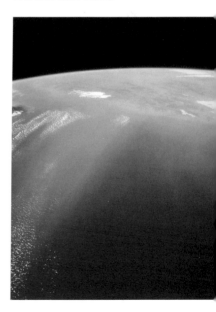

must know

Thermals
When the sun's
rays heat up the
ground, rising
currents of
spiralling hot air
are formed. These
are known as
'thermals', and
provide the lift
exploited by gliders
and birds of prey
such as vultures
and kestrels.

place. In this zone, the air is heated by the ground, and so temperatures fall as altitude increases. Typically this is from around 17°C to about -52°C (63° to -62°F). Where the troposphere ends at a height of between 8–17 kilometres (5–11 miles), there is a transition zone called the tropopause. This has only 10 per cent of the pressure that is experienced at sea level, and there is little change in temperature with altitude.

The stratosphere

Above the tropopause is a layer called the stratosphere, which is found between 17–50 kilometres (11–31 miles) above the ground. In the main part of this zone there are no significant clouds, however, some of the highest forms can be found in the lower stratosphere – these include cirrus, cirrostratus, and cirrocumulus. One of the characteristics of the stratosphere is that temperatures increase with altitude. This is due to the presence of what is known as the ozone layer, which is mostly concentrated at a height of about 25 kilometres (15.5 miles). Ozone is a very reactive kind of oxygen, and the layer it forms is very important to life on Earth. This is because it absorbs many of the harmful rays emitted by the sun – particularly those at the ultraviolet end of the spectrum. In the process of doing so, the ozone molecules absorb

Noctilucent clouds high (75–90 km) above the Earth's surface. While they look much like cirrus clouds, they are, in fact, much thinner and can only be seen during twilight hours at certain times of year, when the sun is just below the horizon.

heat; this then warms the air, raising the temperatures to a high of around -3°C (27°F). Scientists have been monitoring the thickness of the ozone layer for many years now. There is great concern that human activity, especially the release of pollutants called chlorofluorocarbons (CFCs), is responsible for marked reductions in the ozone layer. This has been particularly noticeable over the poles, where the layer has disappeared altogether. Things are improving, however, as CFC emissions have been reduced and the ozone layers are showing signs of recovery. The stratopause separates the stratosphere from the layer above it.

The mesosphere

Above the stratopause is the mesosphere – this reaches up to a height of around 85 kilometres (53 miles). It is the part of the atmosphere where meteors or 'shooting stars' burn up. As there is no ozone layer to warm the air, the temperatures once again fall with increasing height, and they reach a low of about -93°C (-135°F); this is the coldest point in the Earth's atmosphere. The region from the lower stratosphere to the upper mesosphere – which is referred to by scientists as the 'middle atmosphere', has been examined in detail by the ATLAS Spacelab mission series.

A winter sunset over Barter Island, Alaska. Note how the wind has blown the snow into parallel ridges that lie ENE-WSW.

The thermosphere

The layer above the mesosphere is called the thermosphere – this includes the ionosphere, and forms what scientists refer to as the 'upper atmosphere'. It extends to around 640 kilometres (400 miles) above the ground, and is where the space shuttle orbits the Earth. In this zone the air is very thin indeed, and even though far larger by volume than those below it, the thermosphere only contributes 0.1 per cent of the overall mass of the Earth's atmosphere.

This image of Von Karman vortices was acquired by Landsat 7's Enhanced Thematic Mapper. These spiralling winds were created when the prevailing easterly winds swept across the northern Pacific Ocean and encountered Alaska's Aleutian Islands.

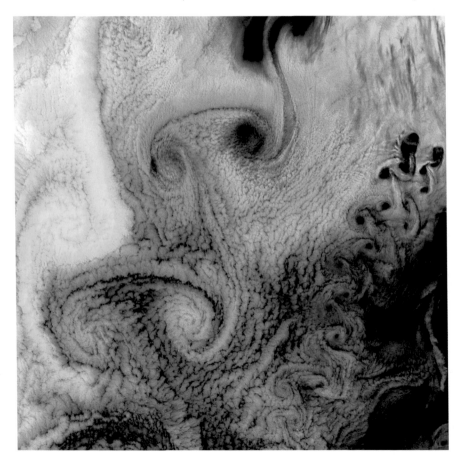

Because the air is so thin, temperatures increase rapidly with altitude as they are heated by solar radiation. They can reach as high as 1,700°C (3,092°F), and as a result chemical reactions between gaseous components happen very quickly. Some molecules in the ionosphere get so hot that they ionize, and this fills the upper parts of the zone with charged particles – these are responsible for reflecting long-distance radio waves back to Earth. If this did not happen, many parts of the world would find it difficult to communicate by radio. The thermosphere is extremely important for life on Earth – it absorbs both 'hard' and 'soft' x-rays, as well as 'extreme ultra-violet radiation', all of which are potentially lethal. Auroras – that is, the Northern and Southern Lights – take place in the ionosphere, and are caused by the charged particles emitted from the Sun by what is known as the 'solar wind'. This is caused by solar processes, and is why there is a direct correlation between auroras and sun spot activity.

The exosphere

Beyond the thermosphere is a layer called the exosphere – it forms the very edge of the Earth's atmosphere, and is where atoms and molecules are lost into outer space. It extends upwards from around 400 miles (640 km) above the ground. Since it graduates into the vacuum of space, there is no specific outer boundary. It is considered by various experts to end anywhere between 1,000 kilometres (625 miles) and 9,600 kilometres (6,000 miles) from the ground, depending on who you talk to. This layer is primarily made up of hydrogen and helium molecules, and as a result the atmosphere is thin enough for satellites to orbit the Earth.

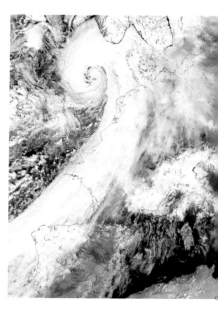

This image was acquired on 30 October 1991, by NASA's Sea-viewing Wide Field-of-view Sensor (SeaWiFS). It shows a storm after it brought torrential rain and hurricane-strength winds to Britain in perhaps the worst event to hit the country since 1987.

Winds of the world

Winds are simply air masses that are moving, although meteorologists usually make a further distinction, and only refer to air that is moving roughly horizontally as wind. Air that moves in a more vertical manner is referred to as a current.

What causes winds?

Winds are caused when masses of air that have different atmospheric pressures flow towards or away from each other. These pressure differences are usually generated by the Sun's radiation heating different areas unequally. In some cases this is because the oceans warm up more slowly than the land, whereas in others it can be due to variations in cloud cover or topographical irregularities. Where such differences in temperature occur, the lighter warm air rises over the heavier cool air, and as a result winds are born. The wind systems of the

A coconut and breadfruit plantation in the Caroline Islands about a week after being hit by a Category 5 typhoon.

world are the result of an amazingly complex series of interactions between such parameters as air temperature, pressure, latitude and direction – even the rotation of the planet has an effect. In simplistic terms, it can be said that hot air in the tropics rises by convection, and then moves towards the poles where it cools and sinks; in doing so, it radiates outwards as dry, icy winds. In reality, however, the picture is much more complicated. There are four major categories of winds – these are the prevailing winds, the seasonal winds, the local winds, and the cyclonic and anticyclonic winds. The last category includes cyclones, hurricanes and tornadoes (see Chapter 5).

Wind zones

There are three main zones of winds – these are the Polar Easterlies, which are found between 60 and 90 degrees of latitude, the Prevailing Westerlies in the 30 to 60 degrees area, and the Tropical Easterlies. These, which are also known as the Trade Winds, are found from the equator to 30 degrees of latitude. Together, these three types form what is known as the 'general circulation' of the world's surface winds.

A distribution map of the world's primary wind belts. Prevailing westerlies are found near the poles, with tropical easterlies either side of the equatorial intertropical convergence zone.

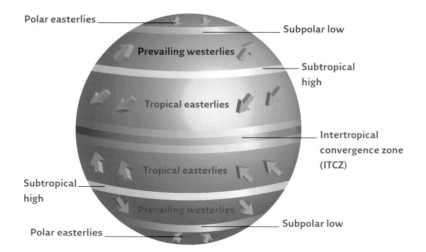

Polar easterlies

Prevailing westerlies

Tropical easterlies

Subpolar low

Subtropical high

Intertropical convergence zone (ITCZ)

Tropical easterlies

Subtropical high

Prevailing westerlies

Polar easterlies

Subpolar low

Weather symbols

In order to make it easier to read weather charts and maps, a series of standard symbols have been established. These may differ between countries, but most are usually easy to understand.

Isobars

Isobars are used to mark lines of constant pressure on a weather map, and are measured in millibars. Other maps of contour lines can be drawn for isotherms, which are points of equal temperature, and isodrosotherms, which are points of equal dew point temperature.

This schematic diagram shows a pair of isobars - the top one is composed of a series of points where the pressure is 996 mb, and the lower one from points at a pressure of 1000 mb.

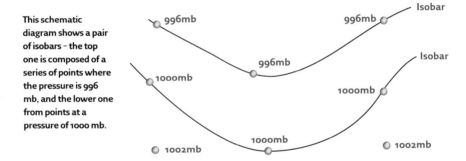

996mb   996mb   Isobar

996mb   Isobar

1000mb   1000mb

1002mb   1000mb   1002mb

Wind vectors

Wind vectors are used on weather maps to indicate the wind direction and its speed. The vectors are represented by arrows which point in the direction of the wind – short ones indicate weak winds, and longer ones show stronger winds.

Geopotential height

Geopotential height is the height of a given atmospheric pressure above sea-level. When a series of these are plotted on a weather map, solid lines called 'height contours' are drawn connecting points of equal height. When small numbers are written alongside the contour lines they indicate the specific height at that point.

Ridges

When a contour map of geopotential heights shows a series of lines bending strongly towards the relevant pole, the shape is referred to as a ridge – these usually bring warm, dry weather.

Troughs

When a contour map of geopotential heights shows a series of lines bending strongly away from the relevant pole, the shape is referred to as a trough – these usually bring cold air and stormy weather.

Convection

Convection is a process in which there are vertical movements in the atmosphere. It is caused by the sun heating the ground - this in turn heats the air above it, and since warm air is lighter than cold air, it begins to rise.

Convergence

Convergence is a process where large amounts of low-level air enter a region from different directions. Where these masses meet, they have nowhere to go except upwards - this then lifts the layer of air above it.

High pressure centres

High pressure centres are also known as anticyclones, and are the zones of highest pressure in a given area. On weather maps they are indicated by a blue 'H'.

Low pressure centres

Low pressure centres are also known as cyclones, and are the zones of highest pressure in a given area. On weather maps they are indicated by a blue 'L'.

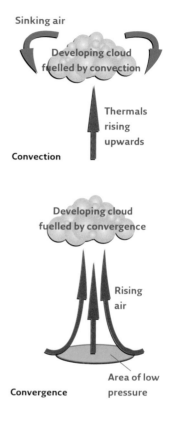

Sinking air

Developing cloud fuelled by convection

Thermals rising upwards

Convection

Developing cloud fuelled by convergence

Rising air

Convergence

Area of low pressure

Fronts

Fronts are simply boundaries between different air masses - they can be cold, warm, occluded or stationary.

Cold fronts

A cold front is the transition zone where warm air is being replaced by cold air. These fronts are denser than warm air, and so when they meet, the cold air pushes the warmer air up. As its altitude increases, the warm air cools and releases any moisture it is carrying as condensation. This creates clouds, which then produce precipitation. When the cooling is particularly sudden, severe thunderstorms can result. On a weather map a cold front is represented by a solid blue line with triangular blocks on the warmer side, indicating its direction of travel.

Warm fronts

A warm front is the transition zone where cold air is being replaced by warm air. As warm fronts progress, they rise up over cold air – this causes condensation and cloud formation, but since they tend to move more slowly than cold fronts, the rate of cooling is lower. The resulting precipitation is therefore milder and more widespread. On a weather map a warm front is represented by a solid red line with semi-circular blocks on the colder side, indicating its direction of travel.

Occluded front

An occluded front is the name given to the situation where a cold front overtakes a warm front – this usually happens as a result of storm activity. It is represented by a solid purple line with alternate

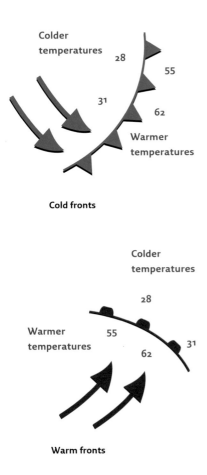

Colder temperatures

28

55

31

62

Warmer temperatures

Cold fronts

Colder temperatures

28

55

31

62

Warmer temperatures

Warm fronts

semi-circular and triangular blocks which point in the direction that the front is moving.

Stationary front
A stationary front is a front that has stopped moving – it is represented on a weather map by a solid line with red semi-circular blocks on the colder side, and blue triangular blocks on the warmer side.

Constant pressure or isobaric surfaces
A 'constant pressure' or 'isobaric surface' is where an atmospheric boundary has the same pressure all along its length.

Measuring wind speeds
Wind speeds are measured on what is known as the Beaufort Wind Scale. This was originally created by the British Rear-Admiral Sir Francis Beaufort in 1805, but has since been amended. It was originally intended to help sailors estimate the wind's speed when they were out at sea, but now uses descriptions of observed effects for use on both land and at sea. The highest sustained speed ever recorded for a wind was 362 km/hr (225 mph) - this measurement was taken on Mount Washington, New Hampshire, United States, on 12 April 1934. It should be noted, however, that much higher speeds occur near the centres of tornadoes.

The Coriolis Force
The rotation of the Earth produces what is known as the 'Coriolis Force'. This effect causes winds to be deflected from their paths and so it has a significant impact on the weather. The magnitude of the effect is determined by several factors, but the most

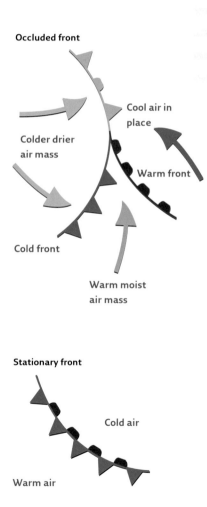

Occluded front

Cool air in place

Colder drier air mass

Warm front

Cold front

Warm moist air mass

Stationary front

Cold air

Warm air

important ones are the speed of the winds and their latitude. The higher the winds' speeds, and the closer they are to the poles, the greater the deflection; at the equator there is no Coriolis Force.

Pressure Gradient Force

The term 'pressure gradient' is used to describe the difference in pressure between two masses of air over a given distance. This difference results in the higher pressure air moving towards the lower pressure area, and results in an overall force that is referred to by meteorologists as the 'Pressure Gradient Force'.

The Beaufort wind scale on land

Beaufort No.	Speed MPH/ Knots	Description	Observed effects
0	0–1/0–1	Calm	Calm; smoke rises vertically.
1	1–3 /1–3	Light Air	Wind direction indicated by smoke drift, but not by wind vanes.
2	4–7/4–6	Light Breeze	Wind felt on face; leaves rustle; vanes moved by wind.
3	8–12/7–10	Gentle Breeze	Leaves and small twigs move constantly; light flags extended.
4	13–18/11–16	Moderate Breeze	Dust and loose paper raised; small branches are moved.
5	19–24/17–21	Fresh Breeze	Small trees in leaf begin to sway.
6	25–31/22–27	Strong Breeze	Large tree branches in motion; whistling heard in telephone cables.
7	32–38/28–33	Near Gale	Whole trees begin to move; inconvenience felt when walking.
8	39–46/34–40	Gale	Twigs and small branches broken off trees.
9	47–54/41–47	Severe Gale	Slight structural damage; large branches broken off trees.
10	55–63/48–55	Storm	Seldom experienced on land; trees uprooted; significant structural damage.
11	64–72/56–63	Violent Storm	Widespread damage.
12	73–83/64–71	Hurricane	Catastrophic damage experienced.

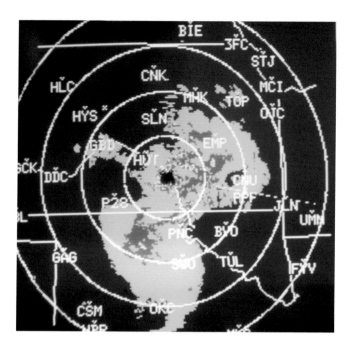

Squally weather passing by the Wichita, USA radar installation and extending from just south of Topeka to Oklahoma City. No violent weather was reported with this particular squall line.

Geostrophic winds

Geostrophic winds are those where the Coriolis Force and the Pressure Gradient Force exactly cancel each other out – this makes them travel parallel to the isobars on a weather map. It is an idealized situation, however, and rarely happens in this exact sense in nature, especially near the ground. Winds begin to form when air masses in a high pressure area start to flow into a low pressure area as a result of the Pressure Gradient Force. As soon as these winds start to move, however, the Coriolis Force starts deflecting them. As the speed continues to increase, the deflection also increases until the Coriolis Force and the Pressure Gradient Force are equal. At this point the winds are said to have become geostrophic. They only occur above the friction layer as the winds within it are influenced by the shape and form of the Earth's surface.

Sunrise at a lake on a misty morning. There is no prevailing wind, so barely a ripple shows on the surface of the water.

Gradient winds

Gradient winds also blow parallel to the isobars on a weather map. They result from the fact that isobars are rarely straight – any curvature introduces a third force known as the centrifugal effect. This is the product of a basic law of physics that states that any moving body left to its own devices will travel in a straight line. A body that is moving on a curved path will therefore always attempt to go straight on; this apparent force is what is known as the centrifugal effect. Its involvement with the Coriolis Force and the Pressure Gradient Force result in the complicated interactions that produce gradient winds.

Surface winds

Winds that blow near the ground are referred to as surface winds. Their proximity with the various features on the ground creates friction – this drag can change the wind's direction, or simply slow it down. At an altitude of around one mile, the ground has virtually no frictional effect on the wind. As a result, the higher layers may well end up travelling in altogether different directions and speeds. Meteorologists refer to the depth of air that is subjected to friction as the 'boundary layer'; in this zone the winds are not truly geostrophic. The thickness of the boundary layer varies depending on a number of factors. These include the air temperature, the time of day, the time of year and the kind of terrain the wind is passing over. The nature of the surface also has a direct bearing on the magnitude of the friction exerted on the wind – calm seas, for instance, have little effect, whereas hill ranges and mature forests have a much greater influence. When the wind slows down, it reduces the effect of the Coriolis Force, the pressure gradient force then becomes more significant, and the direction is deflected towards lower pressure.

The prevailing winds

The prevailing winds arise as a result of global patterns of atmospheric circulation, and include the Westerlies, jet streams, Polar Easterlies and trade winds.

The westerlies

The Westerlies are the dominant winds at mid-latitudes – around 50° to 60° north and south – and blow from the west at low levels from sub-tropical high pressure areas to the sub-polar low pressure zones in the east. Although they are usually mild,

A golf course on the Kona coast of Hawaii. The palm trees have been moulded into shapes by the direction of the prevailing trade winds.

Names of local wind systems

Name	Region
Alizé	From central Africa to the Caribbean
Alizé Maritime	Across West central Africa
Bora	From Eastern Europe to Italy
Chinook	Off the Rocky Mountains
Foehn	Off the northern side of the Alps
Gregale	Greece
Harmattan	Across central Africa
Halny	Northern Carpathians
Khamsin	From North Africa to the eastern Mediterranean
Levanter	The Straits of Gibraltar (easterly)
Libeccio	Italy
Marin	From the Mediterranean into France
Mistral	From central France to the Mediterranean
Nor'easter	Eastern United States
Santa Ana winds	Southern California
Sirocco	From North Africa to southern Europe
Tramontane	From the Alps to the Mediterranean
Vendavel	The Straits of Gibraltar (westerly)
Zonda wind	The Andes in Argentina

they are very changeable, and tend to carry a lot of moisture, and so often transport cyclonic precipitation with them, especially to the west-facing coasts of the continents. When they meet cold air from the polar easterlies, they are forced to rise over them – this causes a sudden drop in their temperature, and heavy precipitation can result. When this happens in winter, it is usually responsible for continental snowfall.

The jet streams

Jet streams are fast moving currents of air that are usually found at heights of between 10–20 kilometres (6–12 miles) above the ground. They can move at speeds of up to 547km/hr (340mph), and measure several thousands of miles long and hundreds of miles wide. They are formed in the Westerlies zone by northward flowing low level warm air that is forced to rise over southward flowing cold air that is also near the ground. The currents are deflected westwards by the Coriolis effect, and are strongest when there is a large difference in the relative temperatures of the Arctic and the tropics. For this reason, jet streams are strongest in the winter – as spring passes and summer takes hold, their strength subsides.

The polar easterlies

The Polar Easterlies are irregular cold winds that blow from the high pressure zones of the poles towards the low pressure areas found around 60 degrees of latitude. They vary from weak to strong, and are forced to travel in an easterly direction due to the Coriolis effect. When these winds meet northward-moving, warm, moist air from the Gulf Stream, the combination often results in violent thunderstorms and tornadoes.

Trade winds

The trade winds are a series of steady prevailing winds that are found in the equatorial region. They blow from the high pressure areas of what are known as the 'horse latitudes', towards low pressure areas around the equator. The horse latitudes are found between 30 and 35 degrees of latitude. They are named after the areas in the sub-tropics where Spanish sailing ships taking horses to the West Indies were often becalmed when the light winds stopped blowing. Since the wind could take weeks to reappear, their water supplies would often run dangerously low. When this happened, the crew would be forced to slaughter the horses, or throw them overboard. The Coriolis Force causes the trade winds to blow predominantly from the north-east in the northern hemisphere, and from the south-east in the southern hemisphere. The zone where they meet is known as the 'doldrums' – this is an area of low pressure that is found between 10° North and 10° South.

Strong winds can make navigation difficult at sea, and so lighthouses can be a great help at night. This is the Europa Point Lighthouse at Gibraltar.

The seasonal winds

Some winds change direction from winter to summer; for this reason, they are known as seasonal winds. They are associated with large continental masses where air over the oceans flows in response to the pressure of the air covering nearby land. In the summer, cold high pressure air over the oceans flows in towards warm low pressure air covering the land. In the winter, this situation is reversed, and the winds blow outwards from the land to the ocean. Examples of these seasonal winds are the monsoons of the China Sea and the Indian Ocean.

Local winds

Local winds are similar to seasonal winds, except that instead of changing with the course of the year, they vary in response to local daily fluctuations. In the late spring and summer, for instance, sea breezes are caused during the day when the air over the land becomes warmer than that over the sea. Land breezes are most marked in the late autumn and winter when the seas are often warmer than the land. They happen at night when the air no longer receives heat radiated from the ground – as a result, it then becomes cooler than that over the sea, and begins to flow seaward. In the mornings and evenings, the relative pressures are not so marked, and the winds tend to be much calmer. Daytime sea breezes reach inland for about 50 km (30 miles) or so, and similarly, land breezes extend for about the same distance out to sea in the cool of the night. Other local winds influenced by diurnal temperature variations include those around topographic features such as mountains and valleys.

Katabatic, Foehn and anabatic winds

Originally, the term 'katabatic wind' simply referred to winds that blew downhill off topographical features such as mountains, hills or glaciers. These days, the definition is usually only applied to winds that are colder than the air they flow into,

must know

Whirlwinds
When conditions are right, short-lived whirlwinds can be formed by concentrations of hot air rising as vigorous spiralling columns. These can pick up lightweight debris and carry it to more than 100 metres (330 feet) above the ground.

Weather symbols

Cold front
The leading edge of an advancing colder air mass. Its passage is usually marked by cloud and precipitation, followed by a drop in temperature and/or humidity.

Warm front
The leading edge of an advancing warmer air mass, the passage of which commonly brings cloud and precipitation followed by increasing temperature and/or humidity.

Occluded front (or 'occlusion')
Occlusions form when the cold front of a depression catches up with the warm front, lifting the warm air between the fronts into a narrow wedge above the surface. Occluded fronts bring cloud and precipitation.

Developing cold/warm front (frontogenesis)
Represent a front that is forming due to increase in temperature gradient at the surface.

Weakening cold/warm front (frontolysis)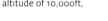
Represents a front that is losing its identity, usually due to rising pressure. Cloud and precipitation becomes increasingly fragmented.

Upper cold/warm front
Upper fronts represent the boundaries between air masses at levels above the surface. For instance, the passage of an upper warm front may bring warmer air at an altitude of 10,000ft, without bringing a change of air mass at the surface.

Quasi-stationary front
A stationary front or slow-moving boundary between two air masses. Cloud and precipitation are usually associated.

Isobars
Contours of equal mean sea-level pressure (MSLP), measured in hectopascals (hPa), MSLP maxima (anticyclones) and minima (depressions) are marked by the letters H (high) and L (Low) on weather charts.

Trough
An elongated area of relatively low surface pressure. The troughs marked on the weather charts may also represent an area of low thickness (thickness trough), or a perturbation in the upper troposphere (upper trough). All are associated with increasing cloud and risk of precipitation.

must know

Wind chill
Although
temperatures
outside may not be
particularly low,
the wind can cool
the body down by
carrying away heat
from exposed
areas of skin. This
is known as 'wind
chill', and it can
make it feel much
colder than it
actually is.

Like clouds, icebergs
can assume fantastic
shapes.

and include the Mistral, Bora and Oroshi winds. Those which carry warmer air are called Foehn winds by meteorologists, and these include the Chinook, Santa Ana and Diablo winds. Those which blow up a slope are called anabatic winds.

Some well-known winds

Many parts of the world experience regular winds for some or all of the year. Examples include the Chinook, the Mistral, the Santa Ana and Sirocco winds. These are discussed here.

Mistral

The Mistral is a very strong, cold north-westerly wind that blows down across southern France and into the Mediterranean Sea during the winter and spring. It is a katabatic wind that is caused by air being cooled over the high ground of the Massif Central and Pyrenean mountains. This then funnels down the Garonne and Rhône valleys, building up speed as it does so. By the time it reaches the coast it is strong enough to cause minor structural damage – it then proceeds out over the Mediterranean where it influences the weather across the North African continent.

Chinook

The Chinook winds, which are named after the local Native American people of the same name, are warm, dry Foehn-type winds. They blow down the eastern slopes of the Rocky Mountains, and across the plains of North America.

Santa Ana

Santa Ana winds are warm and dry, and usually appear in Southern California during autumn and early winter. They are a kind of Foehn wind that is caused by a build-up of air pressure in the high-altitude regions of the Great Basin, between the Sierra Nevada and the Rocky Mountains. When the pressure overflows, it runs down into the adjacent lowlands and out towards the Pacific coast at around 65 km/hr (40 mph). During this journey, it is dried and heated, and as a result the coasts of southern California get their hottest weather while the Santa Ana winds are blowing. There is another, similar down-slope wind that blows just north of the area called the Diablo – this occurs in the San Francisco Bay region.

Sirocco

The Sirocco is a strong wind that originates in the area around the Sahara Desert. It blows across the Mediterranean Sea during the autumn and spring at speeds of up to 100 km/hr (60 mph). Sometimes the Sirocco only lasts a few hours, but it can also go on for days at a time. It stirs up huge amounts of dust, and this, coupled with the high temperatures, is said to be the cause of ill health for people who live along the coasts of northern Africa. It also deposits large amounts of debris in buildings and equipment, and starts storms in the Mediterranean Sea. This wind is responsible for some of the cold, wet weather that is experienced in Europe.

want to know more?

Take it to the next level...

▶ **Hurricanes** 94
▶ **Tornadoes** 104
▶ **Wind speed** 142
▶ **Wind power** 178

Other sources...

▶ Watch the clouds as they cross the sky, and see how the winds that drive them sometimes move in different directions at different heights.
▶ Go and watch a yacht race and see how their skippers cope with changes in the wind.
▶ Examine the structure of wind-borne plant seeds (such as dandelions or old man's beard) and see if you can understand how they are adapted for this method of dispersal.

Weblinks...

For more information about the weather covered in this chapter, visit the following websites:
▶ liftoff.msfc.nasa.gov/academy/space/atmosphere.html
▶ www.usatoday.com/weather/wbarocx.htm

3 Precipitation

Water is an incredible substance – not only is it vital for all forms of life, but it also lies at the heart of the world's weather systems. Some areas have lots of it, whereas others have very little. In many regions, most of it is frozen solid, but in others it only occurs in the gaseous state. The precursors to precipitation are clouds – these form when water vapour condenses into droplets. If these grow in size to the point where they are too heavy for air to hold them, they begin to fall from the cloud as precipitation.

The hydrologic cycle

Precipitation can be in the form of rain, freezing rain, sleet, hail, or snow. Whichever of these it is, though, sooner or later it will continue its part in the endless journey known as the hydrologic cycle. This starts when water falls from the sky as precipitation, and is then later evaporated back into the sky.

Evaporation

Firstly, the heat of the sun causes water to evaporate – this turns it from a liquid state into a gas known as water vapour. The hotter the temperature, the more rapidly evaporation occurs. As water vapour is inherently warm, it rises into the atmosphere, and in doing so, raises the humidity level. Since about 80 per cent of all the water that gets evaporated into the atmosphere comes from the world's seas and oceans, the air above them carries a lot of water vapour. When winds blow this water-laden air over land masses, it can increase the local humidity by a considerable amount. The other 20 per cent of the water evaporated into the atmosphere is derived from fresh water sources such as rivers and lakes, as well as from vegetation.

Red skies can be beautiful to behold. They are usually caused by reflections from minute dust particles that have been trapped high in the air by the atmospheric conditions that precede fine weather.

Condensation

Eventually the air that is carrying the water vapour cools beyond the point at which it can continue to hold it. The water then condenses out as minute droplets; if enough of these are produced, a cloud begins to form. This can happen quite quickly in the right conditions, since warm air will be cooled rapidly if it encounters a mass of cold air. Often, this is as the result of it being elevated by local topographic features such as mountains.

Transportation

The movement of water vapour from over the oceans of the world to the skies above land is known to meteorologists as transportation. While there is a lot of water in the clouds, in reality, most of it is held in the atmosphere as water vapour – in fact, it comprises up to a third of the world's atmospheric gases. Since water vapour is invisible, we cannot tell how much there is in the air merely by looking at the sky. Satellites can, however, determine the amounts, and so it is possible to use the data they produce to generate visual maps of moisture transportation. Clouds, which are moved around under the influence of winds, often dominate the skies, and so are much easier to observe. They are made up of ice crystals or water droplets – or sometimes a mix of both. Although they might appear to contain a large amount of water, it is surprising just how little is actually there; an average cloud that is 1 kilometre high can only produce around 1 millimetre of rain.

Precipitation

The next stage in the hydrologic cycle is the process of water being transferred from the air to the land. This is known as precipitation, and can take the form of rain, hail, sleet or snow. Snow begins as ice crystals which accumulate and then fall towards the ground. Sometimes these pass through air which is warm enough to melt them before they reach the ground. The

must know

A lot of water
Every year around 119,000 cubic kilometres of precipitation falls on land. Of this, about 74,200 cubic kilometres is evaporated back into the atmosphere, leaving 44,800 cubic kilometres as run-off from rivers and streams.

These seemingly endless snow-covered mountains were photographed in Southeast Alaska between Petersburg and Juneau.

Vertically-developed clouds such as these towering cumulus can often develop into severe thunderstorms.

amount of precipitation that occurs varies enormously around the world. In parts of Antarctica, for instance, the temperatures are so low that there is no precipitation at all. Many deserts receive little or no rain for years, whereas the tropical and temperate rainforests may have more than 457cm (180in) per year. Most agricultural areas need around 38cm (15in) annually for crops to grow properly. The amount of rain that falls in a given location is determined by a variety of factors, and periods of drought or floods can bring misery to millions of people.

Groundwater

Once water has reached the ground as the result of precipitation, it is considered by meteorologists to be 'groundwater'. This stage in the cycle concerns the manner in which water soaks into the soil. Nearest the surface is a layer that is called the 'aeration zone', where water and air fill the ground's numerous voids. Below this is the 'saturation zone', where the air is displaced and the gaps are entirely filled with water. The amount of water that a given type of soil can hold is known as the 'porosity', and the speed with which it passes through the soil is called 'permeability'. The line between these two layers is called the water table – this rises and falls depending on the amount of groundwater that is present. When the water table reaches the surface, flooding occurs as there is nowhere for fresh rainfall to go. Likewise, if the permeability is very low – such as when the ground is frozen – flooding is much more likely. When there is sufficient permeability, the water penetrates downwards until it reaches impermeable rock. It then builds up and begins to move sideways; these zones are known as 'aquifers'. Eventually the water carried by aquifers reaches the surface, although the time it takes to do so can vary tremendously, depending on the location. If it gets frozen on its way to the surface by a glacier or by permafrost, it may remain there for thousands of years. Water that does reach the surface then runs downhill until it empties into a stream, river, lake or ocean.

Run-off

Another part of the hydrologic cycle is called 'run-off'. This occurs when precipitation lands on the ground, but does not soak into it. Normally, it drains into a local stream or river, and then makes its way out into the oceans. When there is a larger

Here rain clouds are bringing much-needed precipitation to the parched valleys and highlands of this arid South American landscape.

than usual amount of rainfall, or a sudden thawing of snow, however, the amount of run-off can exceed the capacity of local waterways. When this happens, the end result is, once again, flooding. Floods can be a minor inconvenience, or, as witnessed by millions of people in the New Orleans area of Louisiana in the United States in late 2005, they can be a major disaster, depending on their scale, location, and timing.

Plant transpiration

Not all water returns to the atmosphere by direct evaporation – a significant proportion is transferred by vegetation through what is known as transpiration. This process is responsible for around 10 per cent of the water that is returned to the atmosphere, and so plays an important part in the hydrologic cycle. All plants need water to survive, and most use roots to absorb it from the soil. Those that live in temperate areas are usually able to reach moisture with fairly short root systems, but some desert plants may have roots up to 20 metres (65 feet) long. The method used to pull water up from the ground is very straightforward. Moisture is simply evaporated away from the leaves through special pores – this creates a suction effect that draws further water up the roots and into the main body of the plant.

Flash floods can occur after storms and often wash out roads and bridges, severing communications and leaving entire communities cut off from surrounding areas.

The various forms of precipitation

As we have seen, rain is not the only form of precipitation that occurs – here we examine the others in detail, which include freezing rain, hail, sleet and snow.

This aggregate hailstone made up of several smaller stones has a diameter of approximately 15cm (6in) – the size of a grapefruit.

Freezing rain

Freezing rain is created when snow falls through a layer of air that is just warm enough to melt it into rain. Then, before it reaches the ground, it passes through another layer of air which is much colder. In such situations, the rain drops can become supercooled, where their temperature falls below freezing point but they do not actually freeze. When they hit the ground, however, they immediately freeze – this produces a thin layer of ice which can be exceptionally dangerous. When these conditions occur in areas that do not usually experience such weather, drivers can be caught completely unawares. It is not only a problem on the roads: since the ice it produces can persist for some time, it can also bring down power lines, and cause all manner of other accidents.

Sleet

Sleet is basically made up of partially thawed snowflakes. In different areas the term is used to refer to different forms of precipitation; in western Europe it means a mixture of partially melted snow and cold rain, whereas in the United States sleet is considered to be nearer to hail in form.

The western European form of sleet is more common than freezing rain, but the American form is met with far less frequently. It is very hard to predict when sleet will fall, as it only forms as a result of complex interactions under specific conditions. Sleet can make driving conditions very difficult, with restricted visibility and low levels of grip on the road.

Hail

Hail is similar to sleet in that it is made up of frozen raindrops, but it differs in that it is formed in storm clouds where snow and rain both exist at the same time. Under these conditions snowflakes begin to form as usual, but before they can fall, small water droplets begin to adhere to them. This creates small lumps of ice which then fall towards the bottom of the cloud – before they get there though, updraft currents carry them back up to the top where they accumulate more ice. Under the right conditions the stones can go through this process many times and grow to a considerable size. When they become too heavy to be lifted any more, they fall from the cloud as hailstones. Their size is therefore directly related to the strength of the updraft. Where the stones remain small, the showers usually only last for a short time, and any fallen hail tends to melt very quickly. Hail itself is very visible, so under these conditions it is not as dangerous as freezing rain or sleet. When hailstones reach a significant size, however, they can be a

This fall of large hailstones – many up to 75 mm (3 inches) across – occurred during a severe thunderstorm.

This golf ball-sized hailstone fell near Roosevelt, Oklahoma on 9 April 1978. The individual layers show up as bands of varying opacity.

real threat to both life and property. In rare circumstances, they can be as large as grapefruit. In May 1926, at Dallas, Texas, hailstones as big as baseballs fell from the sky and caused $2 million worth of damage in under 15 minutes. On another occasion, in May 1996, around 100 people were killed, 9,000 injured and 35,000 homes destroyed in China. Indeed, in the United States, apart from severe storms, hail causes more financial loss than any other form of weather with about a billion dollars a year lost in agriculture alone. The heaviest hailstones recorded fell in the Gopalganj district of Bangladesh, on 14 April 1986, and weighed an incredible 1 kg (2.2lb). Their immense size and weight killed more than 90 people; however, it is said that in 1888, hailstones as big as baseballs fell in India. These also killed many unfortunate people who were unable to reach shelter in time.

Snow

Snowflakes are made up of large numbers of minute ice crystals that have joined together as they fall through the air. They have very complicated structures with six-fold symmetry and an intricate, branched pattern which is created by the complex arrangement of the water molecules within them. If the air temperatures are cold enough, the flakes reach the ground without melting. Where this happens, they reach a depth that is approximately ten times higher than that of an equivalent rainfall. In temperate regions, however, the ground is often warm enough to melt the flakes before they begin to accumulate. Light snowfalls are referred to as flurries, but heavier falls are called squalls, and long duration snowstorms known as blizzards. When blizzards occur, they are often accompanied by strong winds. These can blow loose snow for some distance – in doing so visibility can be reduced to zero, and massive drifts created that can make roads impassable and isolate entire communities. The highest recorded fall of snow in a single day was recorded in Silver Lake, Colorado, USA in 1921 with 1.93 metres (76in). It is said

This row of telegraph poles was almost buried by a massive snowstorm that hit Jamestown, North Dakota in March, 1966.

NOAA uses aeroplanes to perform snow surveys.

that on 28 January 1887, a snowflake fell that measured 38cm (15in) across, and was 20cm (8in) thick. The claim was made by a rancher in Fort Keogh, Montana, USA, who said that it was larger than a milk pan. Snow rarely falls at ground level below 35 degrees of latitude since the air temperatures are too high. However, at higher altitudes such as at the tops of mountains, it can form a permanent covering even near the equator; instances of this include Mount Kilimanjaro in Tanzania and in the tropical parts of the Andes.

Floods

When precipitation occurs, there are two main ways in which the resultant water disperses – as discussed above, a certain amount soaks into the ground, and the rest runs away over the surface. If the conditions are suitable, some of this will be evaporated; however, most of it is not. If this results in a build-up of surface water that is beyond the capacity for local run-off, flooding occurs. Depending on the local topography, this may be relatively innocuous, or it may be very dangerous. Where flash flooding occurs, locations that may not even have experienced rainfall can be suddenly deluged with no prior warning. Many hikers and campers have been drowned in this way. Even though

This railroad bridge in Asheville, North Carolina, was washed out by a flood in 1916.

tornadoes are often perceived to be one of the more dangerous natural phenomena, in recent decades they have killed fewer people than flash floods. One of the problems with this kind of flooding is that in most areas they only happen very occasionally. This can lead to apathy – unfortunately, a lot of people who live in susceptible regions do not realize how vulnerable they are. The danger is often exacerbated by modern building practices – covering large areas with concrete or asphalt can significantly reduce the chance for water to soak into the ground. Often storm drains are poorly maintained, and if they get blocked during a downpour, the results can be disastrous. Particularly heavy rainfall can occur if a series of heavy thunderstorms crosses the same ground, or if a single storm is held in one position by local wind conditions. Tropical storms and hurricanes usually deliver the largest amount of rain though. A single storm can deliver over 1 metre (3ft) in a remarkably short period of time.

Thunderstorms

Thunderstorms can evoke all manner of reactions in those who witness them. These range from fear to fascination, for, as with so many weather phenomena, thunderstorms can be both immensely interesting and incredibly dangerous at the same time.

A dangerous phenomenon

The risks include getting caught up in hailstorms, floods, high winds, or being struck by lightning. With the right precautions, none of these should be a problem, but some events defy any preparation. These days we are fortunate enough to know a lot about such natural events – it is not that long ago that they generated enormous fear and confusion. This was certainly the case when in 1638, ball lightning struck the church at Widecombe, in Devon, England. Many worshippers were killed or injured as a result of this terrible incident, as there happened to be a full congregation in the building at the time. The whole place was 'filled with fire and smoke...a great fiery ball came in at the window...'; it was not just a human cost though – the church itself was also 'terribly rent and torn'. At the time no-one could explain what had happened or why, and it was generally accepted that the events were a punishment from God for some unidentified religious infractions.

Types of thunderstorms

Thunderstorms can only occur under certain conditions. There are three basic factors: the air must be sufficiently moist; its temperature must be high enough for unstable updrafts to form; and there must be something to generate lift. This can be in the form of a front, a sea breeze, or it can be caused by local features such as mountains. Although thunderstorms are usually associated with warm or hot summer weather, they can also occur during snowstorms. There are many different types of thunderstorms – these include what are known as single cell, multicell cluster, multicell lines (squalls) and supercells.

Single cell thunderstorms

Single cell thunderstorms are also referred to as 'pulse storms'; these are usually very localized, short-lived affairs that last less than an hour. They are very hard to predict, and tend to form where the vertical wind shear is low; in other words, where there is not much difference in the direction and strength of the wind at different heights. Such storms can be responsible for strong downbursts as

watch out!

What to do if lightning is forecasted

When inside a building
If lightning is predicted, stay indoors, close all windows and doors, and do not be tempted to sit and watch the show. Unplug, and then stay away from all mains-powered electrical appliances.

When away from shelter
Seek shelter if possible – if not, stay away from water and any metallic objects. Avoid open spaces and make sure you are not the tallest object in the vicinity by crouching down if necessary. If you are with other people, do not gather together in a group – instead, spread out so that you are several metres apart. Do not shelter under trees.

This image of cloud-to-ground lightning during a night-time thunderstorm in Norman, Oklahoma, was captured by time-lapse photography.

must know

Even more water
It has been estimated that there are about 1386 million cubic kilometres of free water existing in the atmosphere and on the Earth down to a depth of 2,000 metres below the crust. Of this, about 97.5% is saline, and 2.5% fresh water.

well as hailstorms and even weak tornadoes. They rarely produce severe weather, but can be very dangerous to aviation.

Multicell cluster storms

Multicell cluster storms are the most common form of thunderstorms, and are composed of several individual storm cells which move together. As the storm moves, the combination of gust fronts, atmospheric moisture and unstable downdrafts creates new storm cells around the old ones. At any given time the cells are all at different stages in their own life cycles, and when one of them reaches its peak, it becomes the dominant member of the group until another one takes over. In a long-lasting storm, new cells are created at the upwind edge, mature ones are located in the centre, and the oldest ones are found at the downwind edge. Such events are stronger than single cell storms, but not as powerful as supercell storms. They can produce heavy rainfall, hail, flash floods, and weak tornadoes, and are even more of a danger to aviation than single cell storms.

This photograph taken in February 1984 shows a series of mature thunderstorms gathered near the Parana River in southern Brazil.

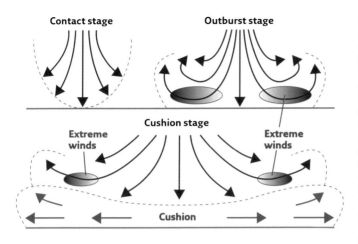

Contact stage

Outburst stage

Cushion stage

Extreme winds

Extreme winds

Cushion

Multicell line storms

As the name suggests, multicell line storms usually form in long lines – they can sometimes be seen in the distance before they arrive, having the appearance of a row of dark clouds. They are also known as squall lines, and although these storms can last for several hours, each cell within them only lasts about half an hour. The most severe weather is generally produced at the leading edge of the storm where the storm's updrafts meet the downdrafts. Sometimes there are severe downbursts, and golf ball sized hailstones can be produced. The main danger occurs, however, when the multicell line stops moving – this causes the thunderstorms within it to keep moving around the same zone. As a result, a large amount of rainfall gets concentrated in a small area, and local drainage systems are likely to be overwhelmed.

Supercell storms

Supercell thunderstorms can produce very severe weather, and as a result are the most dangerous of the four main types. They are typified by a deep rotating updraft that is referred to as a 'mesocyclone'. It is the degree of this rotation that marks the

distinction between multicell and supercell thunderstorms. Such storms are often responsible for heavy rainfall and strong winds; less frequently large hailstones may fall, and sometimes violent tornadoes are spawned. Supercells have been studied intensively by meteorologists in an attempt to achieve better predictions – the precise conditions in which they form are now reasonably well understood, and as a result computer models have improved a lot in recent years. These studies demonstrated that there are three main factors which influence the propagation of supercells – strong winds in the middle and upper troposphere, a high thermal instability, and cross-winds near the ground.

This photograph shows a supercell that developed near Miami, Texas on 19 June 1980. These systems are often associated with violent weather.

Downbursts caused by thunderstorms

When a thunderstorm occurs, high speed low-level winds can be created that are known as downbursts. These are generally split into two forms – these are macrobursts, which are over 4 kilometres (2.5 miles) in diameter, and microbursts, which are under 4 kilometres (2.5 miles) in diameter. Both kinds are particularly dangerous to aviation, but since microbursts are much smaller, they are very hard to detect from any distance. This makes it extremely difficult for pilots who are trying to land to see them in

The out flow from the thunderstorm's core shows up as sheets of wind-driven rain spreading from right to left in this photograph taken in 1982.

time to take avoiding action. If an aircraft encounters a microburst, it is almost impossible for the pilot to be able to retain control as the violent winds change direction very quickly. Under these conditions, it is virtually inevitable that the plane will crash.

Lightning

Lightning is one of the more visible aspects of thunderstorms. It is caused by the movement of air masses that are rising and descending within the storm clouds – this separates out the positive and negative charges, and leaves different areas with different overall charges. The distribution of charge is influenced by several factors, including the

This dramatic image of a thunderstorm over Boston, Massachusetts shows several simultaneous lightning strikes and how they often hit tall buildings and structures.

number of water and ice particles held in the cloud. Eventually, the point is reached where the build-up is no longer sustainable, and a massive electrical discharge occurs in the form of a lightning strike. This takes place either within the cloud itself, or between it and the ground. A lightning strike carries an enormous amount of energy, and this raises the temperature of the air around it to 50,000 degrees Fahrenheit – this is about four times hotter than the surface of the Sun. The result of this extreme heating is that the air expands violently, but milliseconds later – as soon as the strike is over – it cools and contracts again. This causes a massive acoustic shock wave that we refer to as thunder. Since light travels through the air faster than sound, it is possible to estimate the distance of a thunderstorm by counting the interval between the lightning flash and the corresponding peal of thunder. A rule of thumb is that you count the number of seconds and divide them by five to get the distance in miles – for instance, if the period is ten seconds, the storm is two miles away.

Benjamin Franklin's experiment with the kite '...when the string was thoroughly wet, abundance of electricity was procured...'. From: *The Thunder-storm.*

During the course of a thunderstorm, enormous electrical charges build up – this means that a single cloud can produce many lightning strikes, as can be seen here in this dramatic time-lapse photograph.

Lightning is far more dangerous than is generally appreciated – every year it kills more people than tornadoes. It is also far more common than most of us realize; for every second of every day, there are between 50 and 100 lightning strikes around the world. This works out at 8,640,000 times a day. Each strike is about 2 to 10 miles long and carries 100 million volts at a current of 10,000 amps. To put this into perspective, a current of 0.1 amps at 240 volts is more than enough to kill an adult. Having said that, the chances of being struck are only around 1 in 600,000, and even these odds can be reduced significantly by following the simple precautions shown in the 'Watch out!' box on page 69. It is generally accepted by meteorologists that the danger zone around a thunderstorm is about six miles – this, however, includes a large area that is well outside the rainfall zone. As a result, many people are caught unawares, and this is why it is especially important to listen out for weather warnings. A side-effect of lightning is that it can start forest fires that are in themselves extremely dangerous; it is thought that around 15,000 such fires have been caused by lightning in the USA in the last ten years.

want to know more?

Take it to the next level...

▶ Forecasting precipitation 157
▶ Parameters 142-7
▶ Forecasting severe storms 140-1
▶ Acid rain 174

Other sources...
▶ Go and examine the stalagmites in a local cave and see if you can grasp the considerable amounts of time they have taken to form.
▶ Next time you are near a stream or river, have a look to see if you can find the height that the waters reached after the last storm.

Weblinks...
▶ http://observe.arc.nasa.gov/nasa/earth/hydrocycle/hydro1.html
▶ http://www.usgs.gov/themes/flood.html
▶ www.nssl.noaa.gov

OPPOSITE: A lightning strike usually lasts less than a quarter of a second, and is made up of several separate discharges.

A lightning strike has left a small hole in the nose of this NOAA C-130 weather research aircraft.

4 Clouds

Clouds are one of the more visible aspects
of the weather, and are the result of the
evaporation and condensation of water. They
are composed of countless numbers of minute
water droplets or ice crystals, and begin to form
when the humidity level rises to 100 per cent.
The evaporation that starts this process occurs
when water is changed from the liquid state
to its gaseous form by energy from the sun.

How clouds are formed

Clouds are formed as the result of atmospheric water vapour being cooled to the point where it condenses into countless minute water droplets or ice crystals. When enough of these gather together, a cloud gradually becomes visible.

Condensation, convection and convergence

When the air the moisture is carried in cools beyond a certain point, it is unable to hold the water vapour any more, and so this condenses back into a liquid. The process usually happens when warm air rises into cooler air – this can be due to lifting over elevated features on the ground, convection in unstable air, cyclonic convergence, or because it is forced up by frontal lifting. As condensation takes place, millions of tiny water droplets accumulate and a cloud starts to form.

When the sky is clear, a large proportion of the Sun's rays reach the ground, and it begins to heat up. When clouds get in the way, however, much of the energy is instead reflected back

A true-colour image of clouds over the south-western Black Sea on 3 May 2004. The thick white clouds in the centre were caught in a swirling air current while those to the right are wispy, higher altitude clouds that were moving independently of the lower ones.

These menacing gust front clouds mark the boundary between a thunderstorm's cold downdraft and a mass of warm, humid surface air.

into space. Consequently, the amount that reaches the surface is reduced considerably, and temperatures fall. The opposite effect happens at night – when there is a layer of clouds in the sky, they act as an insulating blanket, and temperatures are stable. Conversely, if there is a lack of cloud cover, heat stored in the ground radiates away and it can get cold very quickly.

Cloud classification

Although they come in all manner of shapes and sizes, clouds can be grouped into specific categories. The simplest system has three basic groups – these have been given the Latin names Cirrus, Cumulus and Stratus, which translate respectively as 'curl of hair', 'heap' and 'layer'. Within these groupings there are many further variations. For instance, when the prefix 'cirr' or 'cirro' is added, it means that the cloud forms at very high altitudes – for example, cirrostratus. Similarly, 'alto' (which means 'high') is used to signify clouds that form at medium to high altitudes – for example, altostratus. The term 'cumulo' is sometimes used to signify particular heap-type clouds such as cumulonimbus.

Low-level clouds

Low-level clouds are those where the base lies at less than around 2,000 metres (6,500 feet). They are predominantly made up of water droplets; however, when it is cold enough, these may freeze and turn into ice particles or snow.

Many polar environments are vulnerable as the result of increasing global temperatures – this is the Yahtse Glacier, in Icy Bay, South Central Alaska.

Characteristics of low-level clouds

Certain kinds of low-level clouds can form up in continuous masses, resulting in completely overcast conditions, or they can exist as individual examples in an otherwise clear sky. There are several different types of low-level clouds, including stratus, nimbostratus and stratocumulus, all of which are examined in detail below.

Stratus clouds

Stratus clouds are low-lying white or grey clouds that have no discernible structure – this makes the sky appear flat and opaque. They usually form when there are stable atmospheric conditions, such as when moist 'upslope' air is blown against mountains. One of the better known forms of stratus clouds is fog, which is discussed in more detail later in this chapter.

Stratocumulus clouds

Stratocumulus clouds usually only produce light precipitation – they generally appear as layered, but uneven, rolling masses that can be coloured anywhere from light to dark grey. These clouds are common in coastal areas, and often result in overcast conditions.

Cumulus clouds

When people speak of clouds that look like cotton wool, they are referring to cumulus clouds. These common formations are a function of warm thermals rising from the ground and are usually seen during periods of fair weather. They do not rise to any great extent, and only exist for short periods of time – lasting for anywhere from 5 to 40 minutes. Young ones have distinct outlines, whereas those which are nearing the end of their brief lives have less clearly marked edges. Although they are seen during fine weather, they can sometimes develop into cumulonimbus clouds which can deliver severe storms.

Nimbostratus clouds

Nimbostratus clouds are dark layered rain clouds that produce light or moderate precipitation.

Cumulonimbus clouds

Cumulonimbus clouds are much larger than cumulus clouds, often rising to great heights. When they do this as individual clouds, they are referred to as towers. When they develop side by side, the resulting formation is known as a 'squall line'. They are generated by the action of unstable air – through thermal convection or frontal lifting – which then displaces large amounts of cold air downwards. They usually develop during warm, humid weather, and can produce severe weather including heavy rain, hail, thunder, lightning, strong winds, and tornadoes. Although they are classed as low-level clouds, they can extend to great heights – in excess of 12,000 meters (39,000 feet), which is even beyond the tropopause.

When low-level clouds are viewed from 41,000 feet, it can be seen just how close they are to the ground, and how thin the layer they form really is.

Mid-level clouds

Clouds are classed as mid-level when they form with bases above 2,000 metres (6,500 feet) but below 6,000 metres (20,000 feet). Examples include altocumulus and altostratus clouds.

OPPOSITE: **Here it can be seen just how low-level and mid-level clouds can form as separate layers with a zone of clear air between them.**

Altocumulus clouds

Altocumulus clouds are similar to cumulus clouds, except that they lie at a higher altitude. They often form in parallel bands or round masses as a result of the slow lifting of air ahead of a cold front – they are therefore an indication that rain may well be on its way. If they are seen when the air is warm and humid, thunderstorms may also develop.

Altostratus clouds

Altostratus clouds are similar to stratus clouds, and as such have little or no discernible structure. Their grey colouration often results in sunlight being diffused, creating a translucent sky. They are so-named because 'alto' means mid-height, and 'stratus' means thin, spread-out or layered. They form when a mass of warm air meets a front of cold, dry air.

Altocumulus clouds – here dramatically lit by the setting sun, are mostly made up of water droplets, and lie at heights of between 2,000 metres (6,500 feet) and 6,000 metres (20,000 feet).

High-level clouds

High-level clouds have the prefix 'cirro', and are those which form with bases above 6,000 metres (20,000 feet). At these heights the temperatures are very low – consequently, they are mostly made up of tiny ice crystals.

These high-level cirrus clouds are typically thin and wispy and are known as 'mare's tails'.

What distinguishes high-level clouds?

High-level clouds generally have a thin and wispy appearance, with the better-known examples including cirrus and cirrostratus. When the atmospheric conditions are right, these high-level clouds can produce beautiful optical effects such as halos, iridescence and coronas.

Cirrus clouds

Cirrus clouds are common, high-level clouds that are mostly associated with fair weather. They are usually white, and often hair-like – this stretched-out appearance is due to the ice crystals from which they are composed getting pulled across the sky by the wind. Sometimes the ends of the clouds are curled over, and this has led to them being nicknamed 'mare's tails'. It is believed that cirrus clouds play a significant role in the planet's temperature regulation. There are, however, two directly opposing schools of thought as to what this is. One view is that they are responsible for reflecting significant amounts of the Sun's energy back out into space. If this is so, then they clearly play an important part in helping prevent global warming. It would therefore be ironic that cirrus clouds are often formed by the condensation trails left behind jet aircraft, since air travel is often cited as a primary cause of the Earth's increasing temperatures. The alternative hypothesis is that, in fact, cirrus clouds actually trap heat and contribute to global warming. Until further research is done, we will not know for certain which perspective – if either – is correct.

Cirrostratus clouds

Cirrostratus clouds are more or less the same as stratus clouds, except that they lie at much higher altitudes. They have a dispersed, sheet-like form, and although they are almost transparent, they can be several thousand feet thick. They often develop as the result of a large mass of air being lifted by wide-scale convergence. This causes atmospheric moisture to freeze – the clouds then being made up of huge numbers of tiny ice crystals. Under the right conditions, these can produce spectacular coloured halos and other optical effects.

When the air is humid enough, trails of condensation can form at the wingtips of aircraft.

Contrails

Condensation trails, or 'contrails' for short, are the long white lines that often cross the sky behind jet aircraft. These cirrus-like clouds form at high altitudes from the water vapour that is released by jet engine combustion when the air is sufficiently moist. They are made up of myriads of tiny ice crystals – if the air is too dry, however, then no trail develops. The amount of atmospheric water vapour also determines how long the contrail will last – some disperse very quickly, whereas others can persist for hours.

The white streaks that are so characteristic of cirrus clouds are made by large numbers of ice crystals or snowflakes falling and then being blown sideways by strong winds.

Multi-level clouds

Multi-level clouds do not fit into any of the three previous categories, and include billow, pileus, mammatus and orographic clouds which form in a variety of conditions.

Orographic clouds

Orographic clouds are formed under certain atmospheric conditions when air masses are forced to lift over elevated geographic features such as mountains. As the air rises it cools down, and if the temperature falls low enough, any water vapour it is carrying condenses and a cloud is formed.

Mammatus clouds

The name 'mammatus' is given to small pouch-like clouds which hang under much larger ones – usually these are either cumulus or cumulonimbus. They are usually a blue-grey colour; however, if they are lit by the sun as it sets, they can take on a reddish hue. They are a function of atmospheric turbulence, and although their appearance has been said to indicate that a tornado may be about to form, this is a misconception, and they are in fact, completely harmless.

These mammatus clouds are being lit by the setting sun – their red colouration is due to light being reflected off dust particles suspended in the atmosphere at high altitudes.

Pileus clouds

Pileus clouds are small clouds that form when moist air is lifted by other clouds. They often take on the shape of a cap, and usually lie above or attached to the tops of other clouds – these may be of cumulus, cumulonimbus or stratocumulus form. Alternatively, they may be seen at the tops of mountains.

Billow clouds

Billow clouds – which are also known as wave or 'undulatus' clouds – form in undulating layers or sheets as a result of an atmospheric condition called Kelvin-Helmholtz instability. This phenomenon is similar to the conditions that cause flags to flap. These clouds are so-named because they closely resemble ocean waves breaking on the shore.

This photograph shows a sharp mountain peak in south-east Alaska, on top of which is a small cloud that was formed by orographic lifting of the air.

Fog

Fog is simply a cloud that has its base at very low level. It is caused by moisture from the ground being evaporated – as this rises, it cools and condenses into fog. When this develops into dense banks, it can reduce visibility to practically nothing.

The various forms of fog

Fog can be a severe hazard to everything from shipping to aviation. While ships and planes can use radar to navigate, automobiles cannot, and consequently many people are killed on the roads every year as a direct result of fog. It is therefore important to be able to predict when it is likely to form, and meteorological offices go to a lot of effort to produce detailed fog forecasts. There are two related atmospheric conditions – these are smog and mist. There are several different forms of fog:

This image shows a layer of smog extending across central New York, western Lake Erie and Ohio. This massive cloud of pollution was unable to disperse due to the layer of clouds above it, known as an atmospheric inversion.

Radiation fog

Radiation fog is most common in the autumn, and forms after dark as the ground cools – the conditions have to be right, however. The air has to be carrying sufficient moisture, it must be still, and there must be clear skies. Usually, the fog forms in a low layer that may be less than a metre thick – as a result it disperses quickly when the air is warmed as the Sun rises.

Advection fog

Advection fog is caused by wet air being cooled by low surface temperatures – it often develops over the oceans when warm fronts pass over cold waters.

Steam fog

Steam fog is most commonly seen in low temperature regions, such as those found at high altitudes or around the poles. It occurs when very cold air blows over warmer water, and can be seen over deep lakes during the autumn and winter months.

Precipitation fog

Precipitation fog occurs when precipitation falls through relatively dry air – before it reaches the ground, however, it evaporates into water vapour. When this cools to the dewpoint, it condenses into fog.

Upslope fog

Upslope fog is created in a similar manner to orographic clouds – wind blows moist air uphill, and as it rises it cools. As this happens, the moisture condenses, generating fog.

Valley fog

Valley fog develops in low, enclosed geographical features such as valleys. It is caused by cold air being trapped in the feature by its own weight. When warm air passes over it, moisture condenses out, creating fog. When the winds are light, such fog can persist for several days.

Ice fog

Ice fog occurs when water droplets in the air freeze and turn into minute ice crystals. For this to happen, the air temperatures must be very low – as a result, this kind of fog is usually only seen near the poles.

Freezing fog

Freezing fog develops when water droplets that are suspended in the air become frozen to solid surfaces.

Want to know more?

Take it to the next level...

▶ Condensation 57
▶ Silver lining and cloud iridescence 114
▶ Health and the weather 166
▶ Mankind's activities and the weather 172

Other sources...

▶ Watch the clouds and see if you can tell what the weather is going to do by their shape, size, type or colour.
▶ If you take a flight in an aircraft, look at the cloud formations from above.
▶ For more information about clouds, buy a copy of *The Stories Clouds Tell* by Margaret A. LeMone. (Published by American Meteorological Society, 1993).

Weblinks...

▶ http://www.usatoday.com/weather/wcloudo.htm
▶ www.bbc.co.uk/weather/features/understanding/fog.shtml
▶ www.portfolio.mvm.ed.ac.uk/studentwebs/session4/27/greatsmog52.htm
▶ www.aqmd.gov/smog/

5 Hurricanes and tornadoes

Hurricanes are severe tropical storms that over the years have caused huge amounts of damage and loss of life. They form in the southern Atlantic Ocean, Caribbean Sea, Gulf of Mexico and in the eastern Pacific Ocean. Tornadoes are similar to hurricanes in that they are composed of rotating air; however, they are much more concentrated and, as a result, can be extremely violent – capable of throwing cars, trucks and people high into the air.

Hurricanes

Hurricanes are severe tropical storms that often leave a trail of destruction behind them. They do not just appear from out of nowhere, however, but start out as swirling tropical depressions that grow in intensity until they are categorized as storms.

How hurricanes originate

Severe storms are powered by the release of latent heat when moisture condenses out of cooling air. In this diagram, the converging air in a low pressure system (left) is forcing the air in the centre to rise. As it does, air cools and condensation occurs. This releases latent heat, which expands and draws up more air (right), creating a self-propagating system that can become a hurricane.

A hurricane can develop in a matter of twelve hours or so, when the wind speed rises above 35 knots (62kph/39mph). At this time the rotational structure becomes recognizable on weather charts and satellite images, and the meteorologists assign the system a name. If the atmospheric conditions are right, the storm will continue to build in severity, and may eventually develop into a hurricane. Fortunately, not all storms increase in strength – even so, the heavy rainfall they deliver can bring misery and sometimes death to those exposed to them.

Hurricane classification

A tropical storm is considered to have developed into a hurricane when the average wind speed exceeds 74mph

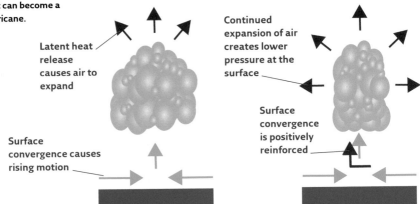

Latent heat release causes air to expand

Surface convergence causes rising motion

Continued expansion of air creates lower pressure at the surface

Surface convergence is positively reinforced

The view to the south-east from a NOAA aeroplane as it passed near the eye of Hurricane Emmy on 25 August 1976.

(64 knots). They are referred to by different names in different parts of the world; the word hurricane is usually only used for severe storms that form over the Atlantic or Eastern Pacific Oceans. Those that develop in the Indian or South Pacific Oceans are generally referred to as 'cyclones', while the ones that appear over the western North Pacific or the Philippines are normally called 'typhoons'. Although most of the hurricanes that are featured in the media occur in the Atlantic Ocean, the vast majority of these severe tropical storms actually take place elsewhere. On average, nearly 100 happen every year, with more than 25 of these forming in the Western North Pacific Ocean.

The region with the highest concentration of hurricanes in the world is the Indian Ocean. This is due to the large number of thunderstorms that form along the Inter-Tropical Convergence Zone – when these combine with the area's warm waters, severe storms often form. While hurricanes impart an enormous human toll, they can also devastate livelihoods and ruin entire economies, such is their ferocity and so great the scope of their destruction. Over the course of the 20th century, for instance, there have been twenty-three hurricanes that have each caused more than $1 billion worth of damage.

Hurricane formation

The exact circumstances in which hurricanes form are not fully understood – this is why they cannot be predicted with any certainty yet. It is known, however, that they only form over the sea, where the surface temperature must be over 26.5°C (81°F). One of the more generally accepted theories is called CISK, or

The five deadliest Atlantic hurricanes of the 20th century

Hurricane Mitch

Hurricane Mitch, which struck Central America in late October 1998, was the fourth strongest hurricane ever recorded in the Atlantic. It produced heavy rainfall and sustained winds of up to 180 mph, which caused floods and mudslides that killed 11,000 people and left 2.5 million homeless. In early November, it then moved into the southern Gulf of Mexico where the warm waters reinvigorated it. The storm then moved over southern Florida, wreaking further havoc.

Galveston, Texas

On the night of 8 September 1900, a Category 4 hurricane hit the town of Galveston, Texas. It was the worst disaster in American history, with more than 8,000 people being killed overnight. Many of these innocent victims had gone to the area on vacation, and did not follow the official warnings to move to higher ground.

Hurricane Fifi

Hurricane Fifi was one of the most devastating storms ever to develop in the Atlantic. It struck in mid-September 1974, when it skimmed the northern coast of Honduras before hitting Belize, Guatemala and Mexico; it then continued into the Pacific Ocean. It is thought that around 8,000 people were killed, mostly as the result of flooding; the total economic cost was about $5 billion.

Hurricane San Zenon

A Category 3 hurricane hit the south-central coast of the Dominican Republic on 3 September 1930. Estimates suggest that around 8,000 people were killed. The storm did immense damage to the town of Santo Domingo, but left others nearby unscathed. It then moved off, crossing Cuba before eventually reaching Florida, where it dissipated.

Hurricane Flora

Hurricane Flora tore through the Caribbean in September and October 1963, causing severe damage in Haiti, Cuba, Tobago and the Dominican Republic. It killed an estimated 7,200 people, and destroyed the livelihoods of hundreds of thousands of poverty-stricken farmers.

'Convective Instability of the Second Kind'. In this hypothesis, hurricanes are considered to be the result of a closed-loop feedback system where the storm cycle continually strengthens itself by releasing latent heat from the atmosphere. Whether this is correct or not, it is certainly the combination of heat and humidity that provide the enormous energy that hurricanes possess. It is also necessary for the atmosphere in the lower and middle troposphere to have high humidity levels. Once hurricanes start to pass over land or cold waters, they immediately begin to weaken as there is insufficient heat or moisture for them to persist. If they stay close to warm water, however, they can last for as long as two, or even three weeks.

Hurricane structure

When hurricanes are viewed from above it can be seen that the storm system is made up of a massive rotating formation with a well-defined spot at the centre. Around this are dense cloud formations that radiate out to the edges of the storm – these are known as 'spiral rain bands'. Due to the Coriolis effect, hurricanes rotate anti-clockwise in the northern hemisphere, and clockwise in the southern hemisphere. The zone at the centre is known as the 'eye' – this is the point about which the hurricane rotates.

To improve hurricane forecasts, meteorologists need to know as much as they can about conditions inside these severe storms. Here, a NOAA research aeroplane is flying in the eye of Hurricane Caroline.

The Saffir-Simpson Hurricane Scale (SSHS)

Hurricanes are either categorized according to the wind speed or the amount of storm surge. A convenient method of assessing a storm's severity was developed in 1969 by Herbert Saffir and Bob Simpson. Called the 'Saffir-Simpson Hurricane Scale', it is often used to estimate the amount of damage that is likely to be caused, and is therefore very helpful in producing accurate advice for the public. It is, however, only used for storms that develop in the Atlantic Ocean and the eastern parts of the northern Pacific Ocean. Other parts of the world use categorization systems which are based on different factors. The method used in Australia, for instance, goes by the peak wind speeds, as opposed to the Saffir-

The Saffir/Simpson hurricane scale (SSHS)

1: Winds: 119-153km/h (74-95mph) or storm surge 1.2-1.5m (4-5ft) above normal
This category of hurricane has little effect on buildings, but there can be extensive damage to mobile homes and trees. There may also be coastal road flooding and minor pier damage.

2: Winds: 155-177km/h (96-110mph) or storm surge 1.8-2.4m (6-8ft) above normal
Hurricanes in this category only cause minor damage to buildings, but can destroy mobile homes and piers, and pull small boats from their moorings. Vegetation can be severely damaged, and coastal areas flooded.

3: Winds: 179-209km/h (111-130mph) or storm surge 2.7-3.6m (9-12ft) above normal
These hurricanes damage small well-built houses, mobile homes will be destroyed and large-scale coastal flooding experienced.

4: Winds: 211-250km/h (131-155mph) or storm surge 4-5.5m (13-18ft) above normal
These powerful hurricanes can cause major damage to well-built houses, and entire roofs will be torn off. Severe coastal flooding and significant erosion of beaches can be experienced.

5: Winds: greater than 250km/h (155mph) or storm surge greater than 5.5m (18ft) above normal
These are the strongest hurricanes of all – even industrial buildings can have their roofs torn off; smaller structures can be completely destroyed. In the event that such a hurricane makes landfall, wide-scale evacuation is necessary for as much as 16km (10 miles) inland.

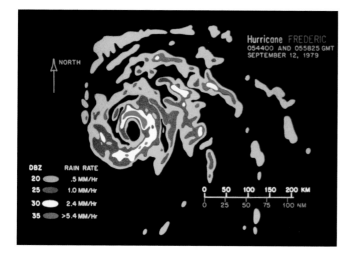

Hurricane FREDERIC
054400 AND 055825 GMT
SEPTEMBER 12, 1979

NORTH

DBZ	RAIN RATE
20	.5 MM/Hr
25	1.0 MM/Hr
30	2.4 MM/Hr
35	>5.4 MM/Hr

0   50   100 150 200 KM
0   25   50   75   100 NM

A coloured radar display of the spiral centre of Hurricane Fred, taken on 12 September 1979.

Simpson scale which uses the maximum sustained wind speeds. While the distinctions between such systems may seem trivial, it is important to bear them in mind when comparing the severity of storms across the world.

Storm surge

Although hurricanes are characterized by high winds, there are many other ways in which they can also present a threat to those in the area. Storm surge, for instance, is the name given to the way the sea level rises in response to the low air pressure and high winds. These also generate high waves, and as a result flooding can be a big problem. A secondary aspect of oceanic waters being pushed up against the shore is that rip tides can develop. These can be a real threat to swimmers, and many unfortunate people have lost their lives to these unseen perils. Rip tides or rip currents are generated when an oceanic swell pushes a large quantity of water onto the shore. As the swell subsides, the water is funnelled between any local underwater obstructions, such as rock formations or sandbars. This can create very powerful currents, which can easily overpower the strongest of swimmers.

Hurricane warnings

At the beginning of the 1900s, many thousands of people were killed or injured by hurricanes as the result of inadequate storm warnings. As weather forecasting has improved, however, it has been possible to issue them much further in advance. Most of this is due to high technology methods such as radar networks and satellite systems, as well as the dedication of many hard-working meteorologists. These days, any atmospheric systems that have the potential to become severe storms are closely monitored by weather forecasters using remote instruments. As a result, hurricanes are usually identified more or less as soon as they begin

Hurricane Norman as it passed south of Cabo San Lucas, Baja California, on 2 September 1978.

to form. If it is thought that a hurricane may make a landfall within 36 hours or so, a hurricane watch is often issued. When this period reduces to 24 hours, the watch is upgraded to a hurricane warning.

Hurricane names

When a tropical depression has increased sufficiently in strength to be categorized as a tropical storm, it is assigned a name. This is done to make it easier to follow individual storms, since several can be in existence at the same time. In the past, several different naming systems have been used. In the West Indies, for instance, hurricanes were given the same name as the saint's day on which they started. In the late 1800s, naval meteorologists used women's names for tropical storms, and this practice was continued when the US National Weather Service began issuing storm names in 1953. Lists

Catastrophic Florida hurricanes: 1900-present

Florida Keys Hurricane	18 October 1906	Category 3 at landfall
Great Miami Hurricane	18 September 1926	Category 4 at landfall
Lake Okeechobee Hurricane	17 September 1928	Category 4 at landfall
Labor Day Hurricane	3 September 1935	Category 5 at landfall
Sanibel Island Hurricane	19 October 1944	Category 3 at landfall
Everglades Hurricane	15 September 1945	Category 4 at landfall
Fort Lauderdale Hurricane	17 September 1947	Category 4 at landfall
Hurricane Easy	5 September 1950	Category 3 at landfall
Hurricane King	18 October 1960	Category 2 at landfall
Hurricane Donna	10 September 1960	Category 4 at landfall
Hurricane Cleo	27 August 1964	Category 2 at landfall
Hurricane Dora	10 September 1964	Category 3 at landfall
Hurricane Eloise	23 September 1975	Category 3 at landfall
Hurricane Andrew	24 August 1992	Category 5 at landfall
Hurricane Opal	4 October 1995	Category 3 at landfall
Hurricane Charley	13 August 2004	Category 4 at landfall
Hurricane Frances	5 September 2004	Category 2 at landfall
Hurricane Ivan	16 September 2004	Category 3 at landfall
Hurricane Jeanne	26 September 2004	Category 3 at landfall
Hurricane Dennis	10 July 2005	Category 3 at landfall

of candidate names are created in alphabetical order, with the exception of those beginning with the letters Q, U and Z. In 1979, male names were added for the first time, on an alternate basis. Whenever a new storm is identified, the next available name is used. There are six lists used for storms that form in the Atlantic – they are used in rotation, so the same names get used every six years. If a particular hurricane is unusually devastating, its name is withdrawn, and is replaced by another name that begins with the same letter.

El Niño and hurricane frequency

El Niño is a phenomenon where there is a sustained sea surface temperature anomaly of more than 0.5°C across the central tropical Pacific Ocean. El Niño years are typically characterized by an increase in the number of tropical storms in the eastern Pacific area, whereas there is usually a decrease in the Atlantic, Gulf of Mexico and the Caribbean Sea regions. Whether this link is direct or merely coincidental has not yet been established by scientists. It is thought that the observed changes in hurricane frequency are due to

The aftermath of Hurricane Camille.

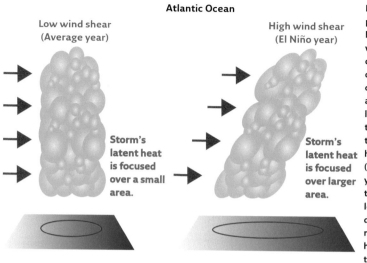

Atlantic Ocean

Low wind shear
(Average year)

High wind shear
(El Niño year)

Storm's latent heat is focused over a small area.

Storm's latent heat is focused over larger area.

Hurricanes are powered by the latent heat released when moisture condenses out of cooling air. When oceanic conditions are 'normal', there is little wind shear, and the concentration of this heat can cause hurricanes to form (left). In an El Niño year, however, it is thought that the high levels of wind shear disrupt this mechanism, and hurricanes are unable to form (right).

differences in the levels of wind shear. The suggestion is that in an El Niño year, there is increased wind shear over the Caribbean and Atlantic. This then disrupts tropical storms, and so prevents them from growing into hurricanes. Conversely, wind shear is reduced in the eastern Pacific, causing an increase in severe storms.

Other factors that may influence hurricane frequency

Although it is known that hurricanes can only form when the surface temperature of the sea is over 26°C (79°F), scientists have not yet managed to establish exactly what happens when the waters are warmer still. This matter is worrying a lot of meteorologists since sea temperatures have risen by an average of 0.50°C (0.9°F) across the world in recent years. It is possible that these rises will result in more storms, or they may last longer or be stronger than previously. Some recent observations suggest that hurricanes do not get any stronger, but that they are more frequent. Until more research has been performed, we will not know for certain what is going on.

Tornadoes

Tornadoes are violently rotating columns of air that hang from cumulonimbus clouds, and are in contact with the ground. They vary in strength from weak, where the wind speed is less than about 175km/h (110mph), to violent where the speeds exceed 320km/h (200mph).

Opposite: Tornadoes are violently rotating columns of air that hang from cumulonimbus clouds. In this example, which hit the town of Cordell, Oklahoma in May 1981, the column is clearly defined.

How tornadoes are formed

Fortunately, the vast majority – about 70 per cent – are in the weakest category. Most tornadoes are formed in severe thunderstorms called supercells where there are violently rotating columns of air. If these grow sufficiently large, a funnel is formed, and the wind speeds increase until a tornado develops. The supercell storms which produce these dangerous conditions are themselves caused by cold air from the poles meeting warm tropical air that is carrying a lot of moisture. Although the United Kingdom has more tornadoes for its size than any other country, they rarely build up to any significant size. In stark contrast is an area in the United States known as 'Tornado Alley', which has the highest concentration of tornadoes in North America. This includes parts of Texas,

In this photograph of a rapidly developing tornado, a large amount of dust and debris has been carried into the air, reaching almost halfway to the cloud base.

The largest measured tornado on record struck near Mulhall, northern Oklahoma, USA on 3 May 1999. It was around 1.6km (1 mile) across, and recordings made via Doppler radar showed that the wind speeds reached a peak of over 480km/h (300mph).

Oklahoma, Kansas and Nebraska, and many of the events the region experiences are classified as strong to violent. Although most tornadoes are produced by the rotations found within supercell thunderstorms, some develop in thunderstorm clouds where there are no such rotational updrafts. These are, however, usually short-lived affairs that do little damage. Every now and then though, they become strong enough to pose a threat, and these rogue storms have been known to kill and injure people. This inconsistency has yet to be explained, and researchers are keen to find out more about this phenomenon. In particular, they want to find out how it is possible for a storm with no rotating updrafts to produce a tornado in the first place. Other tornadoes are formed within hurricanes – these are found in more southern states near the Gulf of Mexico such as Florida, South Carolina and Georgia. Waterspouts or sea spouts are basically tornadoes that form over water. They can form over large rivers and lakes as well as oceans.

The Fujita-Pearson Tornado Scale

One of the ways that the strength of a tornado is measured is by using the Fujita-Pearson Tornado Scale (FPP scale). This was

This tornado which struck Seymour, Texas on 10 April 1979 picked up so much debris that it became completely opaque.

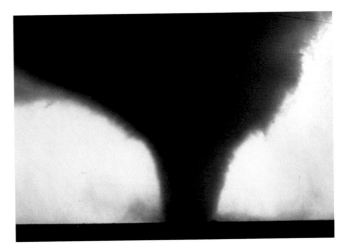

devised in 1971 as a way of determining a tornado's intensity after it has passed by calculating the amount of damage it caused to man-made structures in the area. This method is used because a visual assessment of the size of a tornado's funnel is not a true indication of its strength. The length and width of the tornado's path are also used in the scale, which has six levels which go from Fo, the weakest, to F5 – the strongest. Although other scales exist for measuring the intensity of tornadoes, the Fujita-Pearson scale is the most commonly and widely used in meteorological circles.

The Fujita-Pearson tornado scale

Fo – Gale Tornado (40-72 mph). Path length: 0.3-0.9 miles; Width: 6-17 yards.
This is the weakest category of tornado. It causes light damage to chimneys, branches and sign boards. Shallow-rooted trees may be pushed over.

F1 – Moderate Tornado (73-112 mph). Path length: 1.0-3.1 miles; Width: 18-55 yards.
This category of tornado approaches the wind speed displayed by hurricanes. It causes moderate damage to roofs, mobile homes may be overturned, garages destroyed, and cars pushed off the road.

F2 – Significant Tornado (113-157 mph). Path length: 3.2-9.9 miles; Width: 56-175 yards.
These tornadoes can cause large amounts of damage, with roofs being torn off houses, mobile homes being demolished, railway wagons derailed, large trees snapped or uprooted, and small missiles generated by flying debris.

F3 – Severe Tornado (158-206 mph). Path length: 10-31 miles; Width: 176-566 yards.
These tornadoes cause severe damage, with walls being torn from well-built houses, trains overturned, cars thrown through the air and large numbers of trees uprooted.

F4 – Devastating Tornado (207-260 mph). Path length: 32-99 miles; Width: 0.3-0.9 miles.
In this category, well-built houses can be razed to the ground, and those with weak foundations blown away. Cars can be hurled through the air and large missiles generated from debris.

F5 – Incredible Tornado (261-318 mph). Path length: 100-315 miles; Width: 1.0-3.1 miles.
This is the most severe category of tornado – entire houses are destroyed, cars can be thrown 100 yards or more and even steel reinforced concrete structures can be badly damaged.

Detecting severe storms

Severe storms are the result of strong updrafts forming within storm systems. One way of getting advance warning of their potential strength is therefore to examine the structure of the storm clouds to see if the updraft strength can be calculated.

Meteorological techniques

Storm cloud updraft strength can be assessed by using radar – this provides the data from which visual models of the thunderstorm are generated. There are several different analysis tools used by meteorologists, including what are called the 'Lemon' and 'WRIST' techniques. These work on the following principle – if there is a lot of wind shear, then the updraft gets pushed sideways, but if the updraft itself is strong, then it only gets displaced by a small amount. The solution is thus to identify how much displacement is occurring. Doing this requires a great deal of skill, as the interpretation of the data is far from straightforward, since it has to take into account many complex factors.

The 2005 Atlantic hurricane season

Well before the 2005 Atlantic hurricane season officially began on 1 June, various extended-range forecasts were made. The primary ones were compiled by experts at Colorado State University and the NOAA, and both agreed that it was likely that the number of storms in 2005 would be higher than normal. As the date got closer and more atmospheric recordings were made, these predictions were updated – all of them being revised upwards. In a typical season there are between six and 14 named storms, but by the end of the 2005 season, which ended on 30 November, there had been 23 tropical storms and 27 tropical depressions, making it the most active year on record. Mixed in amongst these were 13 hurricanes – once again, the highest number to date. For the first time ever, the meteorologists actually ran out of official names, and instead had to start using the letters of the Greek alphabet. Three of these storms developed into Category 5 hurricanes – another first for a single season.

The season started earlier than usual, with Hurricanes Dennis and Emily: between them they caused billions of dollars worth of damage. They were eclipsed, however, by Hurricane Katrina. This was the eleventh named storm of the year, and the sixth-strongest ever recorded in the Atlantic basin. It

This dramatic image shows the sun setting in the eye of a hurricane. The red colouration is due to reflections off the huge amount of fine particulate debris in the air.

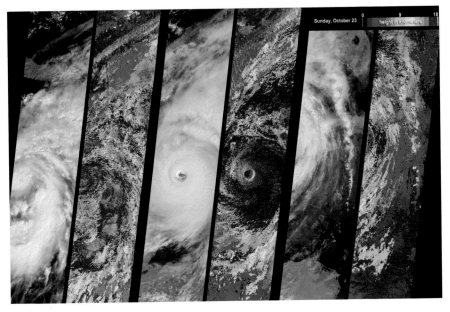

Sunday, October 23

height in kilometers

This spectacular series of images of Hurricane Wilma as it crossed the Caribbean in October 2005 was captured by NASA's Multi-angle Imaging SpectroRadiometer (MISR). In the three pairs of images, the ones on the left show the storm and those on the right show colour-coded images of the heights of the cloud-tops. Purple indicates zero altitude, and red equates to an elevation of 18km (11 miles).

struck south Florida on 25 August while it was still a Category 1 storm. After this, it moved back out to sea where its power continued to develop. On 29 August, it hit the coast of Louisiana – by this time it had become a Category 4 hurricane, and was delivering sustained winds of 235km/h (145mph), which created a powerful storm surge. Tragically, this breached the protective levee system around New Orleans, and huge amounts of water from Lake Pontchartrain and the Mississippi River poured through the breaks and inundated the entire city. The storm also hit the states of Alabama and Mississippi where it caused further damage; to make matters worse, it also spawned at least 36 tornadoes. It is thought that in all over a million people were forced from their homes, and that five million were left without power. Until this episode, the costliest hurricane in American history had occurred in 1992, when Hurricane Andrew struck; Hurricane Katrina though, was far worse. Just as the residents of the region began to repair the damage, Hurricane Rita hit the area; this re-flooded New Orleans, and did further damage to Louisiana and Texas. Other storms later crossed the north-eastern USA, where they also caused severe flooding. It has been estimated that the 2005 season's storms cost well in excess of $130 billion in damage, and that they killed more than 2,800 people.

want to know more?

Take it to the next level...

▶ **Forecasting severe storms** 140–1
▶ **Improving hurricane forecasts** 155–6
▶ **El Niño** 162–4
▶ **Measuring wind speeds** 43

Other sources...
▶ **To find out more about hurricanes, buy a copy of *Hurricane Watch: Forecasting the Deadliest Storms on Earth*, by Dr Bob Sheets & Jack Williams.**
▶ **The next time that a storm passes through your area, see if you can relate its strength to the Saffir/Simpson Hurricane Scale.**

Weblinks...
▶ **www.nhc.noaa.gov/**
▶ **hurricanes.noaa.gov/**
▶ **www.fema.gov/kids/ hurr.htm**

Tropical Storm Beta – the 23rd storm of the 2005 Atlantic hurricane season – formed off the coast of Panama late on 26 October 2005.

6 Unusual and extreme events

This chapter looks at the various unusual and extreme occurrences that are observed around the world from time to time. There is a fine line between what are true 'weather events', and those that are associated with other phenomena. Volcanoes, for instance, are not caused by the weather; they do, however, have a direct effect on it. Likewise, earthquakes are not strictly part of the global weather system. They can cause huge oceanic disturbances though, and so are included here. The Northern and Southern lights are also not the result of our weather, but are a response to conditions on the surface of the Sun. In short, no distinction will be made over the following pages as to what is, and is not, strictly weather-related.

Light effects

Many of the phenomena associated with the weather are the result of optical effects caused by the refraction, dispersion or diffraction of light. These include rainbows, sunsets and haloes, as well as many other well-known visual events. It is worth considering the ways that light behaves in different circumstances.

An amazing optical effect caused by the sun being actually below the horizon during a March sunset.

This Antarctic landscape near the South Pole Station has been made all the more spectacular by the dramatic colours thrown across the sky by the setting sun.

Refraction

Refraction is the name given to the way that light is bent as it passes from one medium to another. The amount of bending is determined by the relative densities of the two media, the angle at which the light strikes, and its wavelength. The longer the wavelength, the less bending occurs. Refraction is responsible for many of the more beautiful optical effects which are seen in the sky. These include haloes around the sun or moon which are caused by light being refracted by ice crystals suspended in cirrostratus clouds.

When white light passes through a prism, it is split into its component colours, and a multi-coloured spectrum is seen – this effect is called dispersion. The separation occurs because in glass different wavelengths of light travel at different speeds. Shorter wavelengths – those at the violet and blue end of the spectrum – are slowed the most, and therefore get bent more than the longer wavelengths such as orange and red. When light is shone past an object, a small amount of it is bent around the edge. The amount of bending depends on several factors, including the light's wavelength. This effect can be seen when light is bent around particles in the atmosphere. This is particularly so with the water droplets that make up clouds, which causes the bright bands often seen along their edges.

Reflection

Reflection is the name given to the way that light bounces off a surface. The rays of light that travel from the source to the object are called incoming, or incident rays, and those that are reflected off it are said to be outgoing, or reflected rays.

Diffraction

Diffraction is the name given to the way that light bends slightly as it passes the edge of an object. The amount of bending is proportional to its wavelength and the gap it is passing between. Where there is a single edge – such as when light shines past the side

A brilliant sunset off the Massachusetts coast in the USA.

of a building, the bending is imperceptible to the naked eye. If, however, it is passing through a small gap between, say, cloud particles, then it may be bent significantly. In this instance, if the conditions are right, there may also be all manner of other optical effects produced, such as fringes of dark, light, or coloured bands. Diffraction is the mechanism responsible for the coronas that sometimes form around the sun or moon, as well as for the silver linings that can occasionally be seen around the edges of clouds.

When the sun is setting, all manner of optical effects can occur. Here, striking rays are the result of the sun shining through low-level clouds.

Scattering

Scattering is the name given to the way that certain atmospheric particles spread light in all directions. This is quite unlike reflection or diffraction, both of which only deflect light in a single direction. Particles that are in the order of 20 micrometers (water droplets) or thereabouts cause 'non-selective' scattering, which makes clouds look white. Light from the Sun enters the cloud and is then scattered in all directions – this is what makes clouds appear white. If the droplets increase in size, or the cloud itself becomes very deep, the light is no longer reflected equally and the cloud appears much darker. 'Mie scattering', on the other hand, is caused by dust, soot and smoke in the troposphere – this makes the sky look hazy. The molecules that make up the atmosphere are very small, and so scatter the shorter wavelengths of light – that is, the ones at the blue end of the spectrum – more effectively than the long ones at the red end. Thus, when sunlight passes through the air, it is selective scattering that makes the sky appear blue. This is known as Rayleigh scattering.

On overcast days, the sun rarely reaches the ground. When there is a gap in the clouds, however, pools of light can form, as seen here.

Optical phenomena

From our perspective on the ground, the Sun shines through the atmosphere from many different angles as it sweeps through the sky on its daily cycle. In doing so, its rays light up the vast number of minute particles suspended in the air in many different ways.

Tricks of the light

Tiny objects in the air that are lit up by the Sun include water droplets, ice crystals, dust, and as we have seen with light scattering, even the air molecules themselves. The resultant interactions produce a wide variety of effects. These include the wonderfully coloured skies which are often seen at sunrise and sunset, crepuscular rays, haloes and coronas, as well as the silver linings and iridescence of clouds.

Haloes are an optical effect where a ring forms around the Sun or moon. They are produced by the refraction of light through a thin layer of ice crystals that lies high in the atmosphere.

Crepuscular rays

When the Sun is rising and setting, it is often partially obscured by clouds or local features, such as mountains or hills. When this happens, its rays fan out across the sky, and if there are sufficient particles in the air, they stand out against the sky in an optical effect known as 'crepuscular rays'.

Water droplets

When the Sun shines through air that contains a large number of water droplets, several different optical effects can occur. Probably the best known of these is the rainbow, but others include cloud iridescence and the silver lining which is sometimes seen along the edge of clouds.

Rainbows

One of the better known optical effects is the rainbow. This is an arc of concentric coloured bands that occurs when the Sun shines

Rainbows are formed by refraction when the Sun shines through air that is carrying large numbers of water droplets. This spectacular example was photographed on the island of Hawaii.

through air that is carrying more than a certain level of water droplets. Typically, this happens when it is raining in one part of the sky and the Sun is shining in another part – the observer has to have the Sun behind them when looking at the rainbow. The colours are formed because the Sun's rays are composed of what is called 'white' light. As these pass through minute drops of water they are refracted. The longer wavelengths at the red end of the spectrum are bent the least, and the blue colours at the other end are bent the most. This causes the primary colours to be separated out in what we see as a rainbow. Although the bands appear to be distinct from one another as seven separate colours – red, orange, yellow, green, blue, indigo and violet, they are, in fact, a whole continuum from one end of the spectrum to the other. Sometimes, two or more rainbows can be seen – this is caused by reflections – and although a secondary rainbow has the same colours as the first (or primary) arc, it has the order reversed. Rainbows are mostly seen only in the summer; this is because they require both rain and sunshine. In winter, the water droplets in clouds are usually frozen into small ice crystals that do not refract light in the same way. If the conditions are right, rainbows can occur by the light of the moon; these lunar rainbows have been recorded since before the times of Aristotle.

Silver lining and cloud iridescence

When the Sun shines from behind a thick cloud, its rays are sometimes diffracted by water droplets to form a silver lining along the edge. This does not always happen, and is dependant on the position of the Sun, and the droplets within the cloud being large enough to cause the effect. A similar optical display can be caused by the same mechanism where the lining is multi-coloured instead of silver. This is known as cloud iridescence, and tends to occur within 20 degrees of the Sun.

Coronas

Coronas around the Sun or moon are formed when light shines through a cloud of water droplets that are of the correct size. The critical factor is that the spacing between the droplets must correspond closely to the wavelength of the light. When this happens, the resultant dispersion and diffraction create concentric rings around the Sun (or moon) – this is called a 'corona'. Occasionally, the corona may appear to have several bands of different colours, with blue on the inside and red on the outside. This effect is due to colour separation, and can only happen if the atmospheric water droplets are very similar in size, and very evenly spaced.

A partial halo with parhelia (sundogs) clearly showing on both sides of the halo.

A parhelion or 'sundog' – also sometimes called a 'mock sun'.

Haloes

There are several different optical effects that are due to the Sun shining through minute ice crystals being suspended in the air – these include haloes, sun pillars and sundogs. A halo occurs when either the Sun or the moon shines through high altitude clouds that are carrying fine suspensions of ice crystals. This creates an apparent ring at 22 degrees from the Sun (or moon) which is usually white, but is sometimes coloured.

Sundogs

Sundogs are an optical effect in which a pair of brightly coloured sun-like spots form on either side of the Sun. They tend to form on cold sunny days when the Sun is near the horizon, either early in the morning or late in the evening. These false suns are caused by the Sun's rays being refracted as they pass through airborne ice crystals. In order for the effect to occur, these must be more than 30 micrometers in diameter, and correctly aligned.

Sun pillars are seen as vertical shafts of light which form either above or below the Sun and are caused by sunlight reflecting off ice crystals falling from the sky.

Sun pillars

Sun pillars are created in similar conditions to those which cause sundogs to appear. They are seen as vertical shafts of light which form either above or below the Sun, and are caused by sunlight reflecting off ice crystals falling from the sky. Sun pillars are most commonly seen when the sun is low at sunrise or sunset.

Sunspots

A sunspot is an area on the surface of the Sun where the temperatures are markedly lower than those of its surroundings. The main area of significance to us as inhabitants of planet Earth is that they generate intense magnetic energy and are linked to increased solar activity.

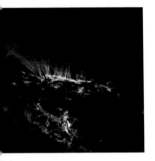

This spectacular ultraviolet image was taken by NASA's sun-observing TRACE spacecraft in September 2000. The gases around the large bright sunspot have a temperature of over one million degrees Celsius (1.8 million degrees Fahrenheit).

The dangers of sunspots

The incidence of sunspots can affect both our weather and much of the technology we rely on. When large sunspots appear, they can knock out communications systems, and in extreme cases, even power networks. However, the extent to which they can affect our climate is still a matter of much debate amongst scientists. Some believe they may be responsible for increases in global temperatures, whereas others disagree strongly. It is known that there has been more sunspot activity in the last 70 years than there has been over the previous 8,000 years; however, a definitive connection to climate change has still to be proven.

Aurora Borealis

When sunspot activity is high, one of the more visible responses here on Earth shows up at the North and South Poles as a magnificent nocturnal light display. These phenomena are known respectively as the Northern Lights or Aurora Borealis, and the Southern Lights or Aurora Australis.

They show up as intense arrays of colour that move around in a blaze of red, green, blue and violet. The ions that create the aurorae are emitted from the sun as bursts of highly charged particles, which are known collectively as plasma. These streams form what is called the solar wind, and when this reaches the outermost parts of the Earth's atmosphere, they interact with it in various ways. Some of the particles are reflected back out into space, but others get caught in the Earth's magnetic field and are directed towards the two poles. Eventually they get close enough to begin colliding with the molecules in the ionosphere, at which point they start glowing. When sunspot activity is high, this can produce spectacular displays in the night-time skies of high latitude regions. The energy involved is immense, with around a million megawatts being consumed by the more vivid aurora. This creates enormous amounts of interference, and is why electrical equipment can be damaged. Scientists are therefore very interested in finding out more about sunspots; although there is still much to discover, it is now possible to predict the aurorae with some accuracy. In the longer term it may be possible to utilize the energy they carry, but until then all we can do is to enjoy the displays whilst trying to minimize the damage they can cause in many different ways.

Here, the Aurora Australis or Southern Lights are seen over Kangaroo Island, South Australia. As with the Aurora Borealis, these are seen during strong geomagnetic events.

Extreme events

Some events are not directly caused by the weather, but can have significant effects on it. Phenomena like wildfires, drought, and sand or dust storms are often simply the product of a lack of precipitation, but things like 'weird rain' often defy explanation.

must know

When a volcano erupts, it can throw large amounts of noxious material into the air. Between 1 and 5 per cent of this is gas of one sort or another, with water vapour being by far the largest component.

OPPOSITE: **This image of Mount St. Helens taken on 7 August, 1980, shows a pyroclastic flow. These powerful events can move at speeds of over 100km/h (60mph) and reach temperatures of over 427°C (800°F).**

Volcanoes

Although volcanoes are not caused by the weather, they can certainly have a marked effect on it. When a typical eruption occurs, it can launch massive quantities of material into the atmosphere. This can block out sunlight over vast areas, and if this happens for any length of time, it can cause global temperatures to fall. One of the largest eruptions of the 20th century took place in 1991 when Mount Pinatubo in the Philippines blew up. Over the following year, the atmospheric disturbance it created caused temperatures across the world to fall by about half a degree on average – an enormous amount in meteorological terms. It is estimated that about 20 million tonnes of sulphur dioxide were released into the upper atmosphere. When this reached the stratosphere it was converted into small droplets of sulphuric acid called aerosols. These became suspended in the air, forming a layer that reflected large amounts of sunlight back into space.

For a fuller picture of what can happen, we have to revert back into history. It is thought that when Mount Tambora erupted in Indonesia in the spring of 1815, it was one of the most powerful volcanic events of the last 10,000 years. The release of ash, gases, and other materials into the atmosphere caused major changes in the weather right across the northern hemisphere. The protracted winter and endless rainstorms led to massive crop failures, and many thousands of people starved to death. As if this was not bad enough, the famine then became a typhus epidemic which killed around 200,000 people in Europe alone.

must know

Consequences

Although
earthquakes are
unconnected with
the weather, they
can trigger similar
catastrophic
results, including
tsunami such as
the one which
killed around
275,000 people on
Boxing Day, 2004.

Earthquakes

The Ancient Greeks thought that earthquakes were caused by
winds that had become trapped in caves below the Earth's
surface, and that until they broke free there would be calm
weather. For this reason, they believed that all earthquakes were
preceded by hot, calm periods. Later on, this idea was changed
to state that sightings of fireballs and meteors were advance
warnings of earthquakes. These days we know that there is no
link at all between earthquakes and the weather as they are
produced by geologic processes deep below the Earth's surface.
They do, however, change the shape of the landscape as they are
a result of the movement of the plates that make up the Earth's
crust. They are therefore responsible for both elevating and
lowering whole areas; this can result in the creation of
everything from mountain ranges to new inland seas. Although
it may seem that earthquakes are more common than they used
to be, this is not borne out by the facts. The misconception is in
reality due to two main factors. Firstly, it is because of better
reporting through vastly improved communications networks –
in the past we were often not aware that an event had occurred.
Secondly, it is because there are now far more seismic recording
stations. In 1931, there were only about 350 such sites. These
days there are about 4,000 spread around the world, from which
there are an average of about 35 earthquakes recorded every day.

Wildfires

Fire has been an instrument of nature since well before mankind
evolved – indeed, there are many species of plants that rely on fire
to open their seed pods. Without it, they would simply not survive.
Much of North America's landscape has been shaped by wildfires,
especially in the Great Lakes region. However, since mankind –
especially the Europeans – has populated the continent, the once
extensive forests have been considerably reduced in size. These
days, many fires are started deliberately – either through malice or

carelessness. Others start when lightning strikes hit dry material; either way, they can wreak havoc over massive areas. One particularly bad fire began in October 1871, and more than 810,000 hectares (2,500,000 acres) were destroyed and 1,500 lives lost. Sadly, many more have occurred since then; in the Great Lakes area alone there are more than 6,000 fires every year. One of the main factors in determining the seriousness of a fire is the weather; when the moisture level is low, forests can become tinderboxes. Under these conditions, the slightest spark can start a wildfire. Once the flames have taken hold, the wind becomes a predominant factor since it governs the fire's speed and direction. Although fire crews are successful in managing to extinguish many of these conflagrations, the bigger ones often only end when the weather changes and enough rain falls to stop them spreading any further. Fires have important links to weather and climate by adding smoke and soot to the atmosphere and changing landcover.

When there is a drought, forests can become tinderboxes, and under these conditions wildfires can wreak havoc over vast areas. Whilst some are caused by lightning, many are deliberately started.

Under normal conditions, this lake in Northern France is fresh and clean. However, when this photograph was taken the whole region was suffering from a drought and the levels had fallen until all that remained was a green sludge.

Weird rain

Of the many weather phenomena, perhaps the most inexplicable are the many accounts of unusual objects, such as frogs and fish falling from the sky. While it is possible that some of these were sucked up into the air by tornadoes or sea spouts and then deposited many miles away, some accounts defy rational thinking. For instance, during a storm thousands of frogs suddenly appeared out of the sky and landed on Kansas City, Missouri, in 1873. A similar event happened in Minneapolis, Minnesota in July 1901. A local news report said that 'When the storm was at its highest... there appeared as if descending directly from the sky a huge green mass. Then followed a peculiar patter, unlike that of rain or hail. When the storm abated the people found, three inches deep and covering an area of more than four blocks, a collection of a most striking variety of frogs... so thick in some places [that] travel was impossible.' While these events are highly unusual, they continue to occur to this day – an illustration of just how far objects can be carried in the air is perhaps best demonstrated by the account of a large number of North African frogs which fell on the people of Naphlion, in southern Greece, in May, 1981.

It is not just frogs falling from the sky that have astonished onlookers. In February, 1861, shortly after an earthquake, a shower of fish fell on parts of Singapore; no-one knows to this day if the two events were linked. Other places that have seen fish-rain have been as far apart as Bournemouth, England, where a shower of herring fell in 1948, and North Sydney, Australia, where, in 1966, a priest was hit on the shoulder by a large fish.

It was reported that a shower of bird's blood rained from the skies in 1890, in Messignadi, Calabria in Italy, although where it came from is not known. Speculation at the time suggested that a flock of birds were torn apart by fierce winds, but this remains unproven. A similar event happened in 1841, when a tobacco farm near Lebanon, Tennessee in the United States was suddenly inundated with blood, fat and muscle tissue. Yet another shower of blood fell on a farm in the Los Nietos Township, California, USA in 1869. Other, less grisly showers have included tons of periwinkles and hermit crabs which fell in Worcester, England in 1881, what appears to have been jellyfish slime in Tasmania in 1996, and several showers of kernels of corn in Colorado, USA between 1982 and 1986. However, the most bizarre of all these events must be the shower of live baby alligators which fell in 1877 on a farm in South Carolina, USA.

Drought

Droughts have occurred on and off in many places on Earth since time immemorial. One of the more general definitions is that they are periods of abnormally dry weather which are bad enough to cause 'serious hydrologic imbalance in the affected area'. In layman's terms, this means that crops fail or there are severe water shortages. They can be brought about by many different environmental factors. For instance, in areas that rely on melt waters, a lack of snow during the winter can cause major problems in the spring and summer. For city dwellers, the onset of a drought may only mean that they are not allowed to wash their cars for a few weeks, but for peasant farmers, it can be a matter of life and death. In many areas, drinking water is supplied from reservoirs that usually fill up enough during the months with high precipitation levels. If there is less than the expected amount, however, the reservoirs can run dry, leaving local inhabitants with significant problems. As a result of this, meteorologists put a lot of effort into trying to improve their drought prediction methods.

One of the better-known periods of extended drought occurred in the 1930s when the Great Plains of the United States turned into what became referred to as the 'Dust Bowls'. In this sad period of American history, some 50,000,000 acres of land were affected by severe water shortages, and crops failed for many years. Farmers throughout the region struggled

This image from the late 1930s of severe land erosion caused by poor farming practices shows what happened during the dust bowl years that hit the United States so badly.

OPPOSITE: In times of drought, entire rivers can be reduced to a mere trickle. While some kinds of wildlife are adapted to cope with these events, many are not and quickly die out.

to survive, and large numbers had to give up and move to the cities to support their families. Although the way the land turned to dust has long been blamed on poor farming practices, recent studies have shown that such events have occurred in the area many times, well before any European settlers arrived.

Sand and dust storms

Sandstorms and dust storms are common in many arid regions that experience high winds; these include the Great Plains of North America, the Sahara Desert of North Africa, the Gobi Desert of Mongolia, the Taklamakan Desert of north-west China and the Thar Desert of India. They are usually caused by updrafts of air that form over hot ground as a result of convection. As the hot air rises, cooler air blows in from the

This image of a terrifying dust storm bearing down on Stratford, Texas dates from 18 April, 1935.

sides to replace it, and any loose material on the ground can be picked up and carried for thousands of miles. Recent scientific evidence has shown that sand from the Sahara contributes valuable minerals to the rainforests of the Amazon, as well as to marine plankton in the western Atlantic Ocean. Where the wind is very powerful, entire sand dunes can be blown away, and in severe storms, the fronts can appear as sheer walls of dust and sand that may be up to 1,525m (5,000ft) high. Conversely, the airborne material is often deposited in places where it is not wanted. In these circumstances, roads and railway tracks can be obliterated, and houses buried. Over the course of history, many entire cities have disappeared beneath dune systems created by sandstorms. In agricultural regions, these storms can remove all the topsoil for miles, leading to severe hardship for local farmers. When such a storm is blowing, visibility is reduced to zero, and travel becomes impossible.

want to know more?

Take it to the next level...

▶ **Hydrologic cycle** 56-61
▶ **Condensation** 80-1
▶ **Lightning** 74-7
▶ **Weather satellites** 151-5

Other sources...
▶ **When you see a rainbow, look carefully to see if there is a secondary arc. If there is, look to see how the colours are reversed.**
▶ **For more detailed information on optical phenomena, buy a copy of** Rainbows, Haloes, and Glories, **Robert Greenler.**

Weblinks...
▶ **www.bbc.co.uk/ weather/features/ understanding/mirages. shtml**
▶ **www.usatoday.com/ weather/wildfire.htm**
▶ **vulcan.wr.usgs.gov/ Glossary/VolcWeather/ description_volcanoes_ and_weather.html**

Dust storms were a regular feature during the dust bowl years of the mid-1930s. They buried houses, farms and even small communities, causing widespread suffering and death.

7 Forecasting

As we have already seen, weather forecasting
has been of great importance to mankind since
the earliest times. We discussed the forerunners
of the subject in Chapter 1, so this chapter will
focus on more recent forecasting methods.
Recent extreme weather events have
demonstrated only too well that there is still a
lot of pressure to improve weather forecasting
techniques. Not only do many people's lives rely
upon this, but also the fate of entire economies.

Basic observation

There are many different ways of determining what the weather is going to do, but some are definitely more reliable than others. This is because weather systems are so complex in nature that it is not possible to make forecasts over extended periods of time with any real degree of certainty.

A co-operative weather station at Granger, Utah which was run by local volunteers to record weather parameters such as temperature, precipitation and sky conditions.

Forecasting starts with observations

Minute changes in atmospheric conditions can build up over a day or so to cause marked differences from those which were expected. In other words, the longer-term the prediction, the less accurate it is likely to be. Where the climate is relatively stable, detailed forecasts can be made for five to six days in advance. In regions where conditions are less stable – for instance, where there are major geographical factors to consider – this can drop to three or four days. Examples of the latter include coastal areas – especially peninsulas, where the synopsis can change very quickly.

All weather forecasting is based on taking measurements of recent conditions, and comparing them with current observations. There are several different ways of using this information to make weather predictions though. One of the simplest concepts is what is known as the 'Persistence Method' – this simply works on the principle that tomorrow's weather is likely to be the same as today's. In other words, if it is snowing today, it will probably snow tomorrow. This way of predicting the weather really only works well in regions where conditions do not change very often. Over longer periods, however, the persistence method compares very well with some of the more advanced techniques, such as numerical modelling.

A second technique called the 'trends method', works by performing large numbers of mathematical calculations to

Another NOAA aeroplane sets off to make atmospheric observations.

identify the way various weather features such as low pressure centres and fronts are likely to change with time. For instance, if a cold front is travelling at a steady 16km/h (10mph), it is likely that in ten hours it will have moved 160km (100 miles). The results of these calculations are then used to establish a predicted outlook. This method works very well over short periods of time, especially if there are few major components to take into account. If the scenario is more complicated, however, it is much less accurate.

These days most meteorologists exploit the power of computers to process all the data they have collected to make their forecasts. This method, which is called 'Numerical Weather Prediction' uses programs that are known as 'forecasting models'. They rely on vast data-sets of atmospheric measurements and require huge amounts of computational processing power. As a result they have to run on purpose-built supercomputers. The main weakness of this method is that the algorithms which drive them are based on scientific principles that are not yet fully understood. Consequently, the models work best when the weather picture is relatively straightforward. Another weakness is that the system also relies on having complete data-sets. In regions where there are few

This is a solar-powered Surface Automated Measurement (SAM) site used to take measurements in and around severe weather.

recording stations, such as in mountainous or oceanic areas, this is often not possible. Such gaps in the data considerably reduce the accuracy of any predictions. Nevertheless, it is still the most popular method used today for short-term forecasting. Meteorologists have found that they can improve their numerical model forecasts by getting the computers to run the data through several times. Since small changes in atmospheric conditions can make big differences to what happens a day or so later, the trick is to vary the starting conditions slightly on each run – if the results are similar, then the forecast is issued. If not, they try again with different parameters. Unfortunately, the inaccessibility and need for enormous

Helicopters are often used to access remote areas.

computational power means that as a technique, numerical modelling is out of reach for all but the most professional of forecasters.

Long-range forecasting

Although forecasters can currently make reasonably detailed predictions for about five or six days in advance, the experts are doubtful that their techniques will ever improve beyond a period of two or three weeks ahead. It is, however, still possible to make outlook forecasts over several months that cover many months. They are usually based on the analysis of statistical information derived from recent measurements – especially those of local oceanic temperatures. This is because deep water is thermally very stable, and as a result it takes a long time to warm up or cool down; it therefore has a significant influence on any winds that blow over it. Such predictions are far less detailed than short-term forecasts – that is, they do not try to specify exactly what will happen on a given day. They can, however, say what the likelihood is that it will be hot

When meteorological studies are being conducted, it is vital that the local topography is well recorded. Here, in Wide Bay, Alaska, a theodolite used for surveying the landscape is silhouetted by the setting sun.

In 1984, a study known as Project Chase was initiated by the National Severe Storms Laboratory to follow and document tornadoes. Researchers scan the horizon for likely candidates.

Doppler radar has been a key weather forecasting tool for many years now. This was what the installation near Norman, Oklahoma, looked like in 1973.

or cold. While forecasters might say, for instance, that a given month in winter might be wet, they are unable to tell whether it will snow or not. This is because snow only forms under certain conditions that are far too exacting for any long-range assessments.

Detecting severe storms

Forecasters have two main responsibilities to the public – not only do they have to provide 'run of the mill' weather predictions, but they are also expected to give early warnings about severe storms that could threaten lives or livelihoods. Working out just when such events are likely to take place is exceedingly difficult, and meteorological researchers devote a lot of time to improving their techniques. Once again, they rely heavily on computer modelling to examine exactly which atmospheric conditions lead to storm development, and which do not. Once these have been identified, the models are run over and over again with very small variations – this allows the severity of the resultant storm simulation to be assessed. Such research has found that one of the main factors that determines a storm's strength is the degree of vertical wind shear – this is the amount of change in the wind's direction and height. Consequently, meteorologists now know what they have to look out for, and the accuracy of storm prediction has improved tremendously.

This is the display produced by the Norman Doppler radar installation of a squall line that developed on 11 June, 1985. The image is produced by computer interpretation of radio waves reflected by atmospheric features.

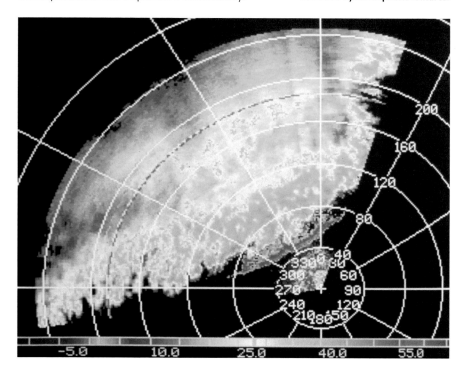

The parameters of forecasting

Weather forecasting relies on the observation and recording of atmospheric conditions across wide areas. While some of these measurements are taken with simple instruments, others are provided by the latest high-technology equipment.

Wind socks
Aircraft pilots need to know what the wind speed and direction is both when taking off and landing. In order for them to be able to tell what the speed is from a distance, a device called the wind sock was developed. This uses a large conical fabric tube that is open at both ends, and secured to a long mounting pole. When it is calm, the brightly-coloured sock droops down vertically, but as the wind blows, it rises until it reaches the horizontal; in doing so, it points away from the wind.

There are many different parameters to consider, and since forecasts can only ever be as accurate as the data they are built from, it is important to understand what these are and how they are represented on weather charts. Examples include the amount of moisture in the air, its temperature, the speed and direction of the wind, and the atmospheric pressure. When these are measured and the data collated, it is possible to start mapping the information. This in turn makes it possible to identify the location and nature of features such as high pressure centres or storm fronts.

Air temperature

This is the temperature of still air and, depending on the local convention, is measured in degrees Fahrenheit or Celsius.
Fahrenheit = Celsius x 1.8 + 32

Wind speed

The average wind speed is measured in knots using a device called an anemometer – one knot is one nautical mile per hour:

1 knot = 1.15 miles per hour = 1.9 kilometres per hour

Wind is usually represented on weather charts as 'wind barb' symbols – these indicate the direction of the wind and its speed. The strength of the wind is shown by the 'prongs' on the barb – if the air is calm, there are none shown, and as the speed increases, more barbs are added. A short barb means a wind speed of 5 knots, and a long barb means 10 knots. Shown together, they indicate a speed of 15 knots; if there are very strong winds, then a pennant is added – this represents a speed of 50 knots. Using this system it is possible to construct a simple, but accurate, map of wind speeds.

Wind direction

The direction of the wind is measured in degrees, and indicates where the wind is coming from. If a

The standard instrument shelter or Stevenson Screen is one of the standard tools of the meteorologist. It houses instruments which record temperature, pressure, and relative humidity.

wind is described as a south-westerly, this means that it is blowing from the south-west, not towards it.

Air pressure

Air pressure is measured in three main ways – some reports are given in inches of mercury ("Hg), others in millibars (mb), and still others in hecto Pascals (hPa). A millibar is 1/1000th of a bar, and is equivalent to 1 hPa. One inch of mercury = 3386.389 Pascals at 0°C.

1 inHg = 3386.389 Pascals at 0°C

1 hPa = 0.001 bar

1000 hPa = 100,000 Pascal = 14.50 pounds per square inch = 750.0 mmHg = 0.9869 atmosphere

Since air pressure varies with altitude, meteorologists use the value measured at sea level as a baseline; the average pressure at sea level is 1013.25 millibars.

Dew point

The 'dew point' is a figure which indicates how much moisture there is in the air. Specifically, it is the temperature to which air must be cooled (at constant pressure) before the moisture it contains condenses into water. If the temperature falls below freezing, it is instead called the 'frost point'. If the relative humidity is high, then the dew point will be close to the temperature of the air. When the relative humidity reaches 100 per cent, the dew point becomes the same as the air temperature.

Humidity

Humidity is a measure of the amount of water vapour that is in the air. There are several specific variations on the term, but the main

OPPOSITE: **This mini weather station is used to control the environment inside a large shopping mall in Paris. It measures the wind speed, direction, temperature, humidity and light levels.**

one used by meteorologists is 'relative humidity'. Specifically, this is the ratio of the amount of water vapour that is in the air, compared to the amount that it is possible for the air to hold under the current conditions. If the air has the maximum amount – about 23 grams per cubic metre (at sea level and at 25°C/77°F), then it is said to be saturated, and the relative humidity figure is 100 per cent. The amount of water that the air can hold is not constant, though – it falls dramatically with temperature: the figure of 23 grams at 25°C/77°F drops to 5 grams at 0°C. A relative humidity figure of around 45 per cent is considered by most people to be the most comfortable level.

Rainfall

The amount of rain that falls is recorded by meteorologists and hydrologists using devices called rain gauges, of which there are many different kinds. Some are designed so that they can also measure snowfall. They work by either measuring the volume or the weight of any precipitation that falls – types include cylinder gauges, weight gauges, tipping bucket gauges and pit collectors.

The graduations used also vary: in some parts of the world they measure in inches, whereas in others they use millimetres. While they tend to be very simple in construction, they are not foolproof – for instance, if the wind speed is high, the gauges can become wildly inaccurate. They can also fail to record true precipitation levels if the collecting funnel gets blocked by ice or snow. Some rain gauges are read frequently by observers – who are often volunteers, whereas others, usually those in remote locations, are read automatically. When

This photograph from the 1930s shows the rain gauge at the US Smelting Co. co-operative weather station in Midvale, Utah.

they are read, the water that has been collected is usually discarded, but if necessary, samples are kept for pollutant testing.

Standard rain gauge

The most common type of rain gauge is simply constructed from a funnel which leads into a graduated cylinder. In areas where high precipitation is expected, this usually has an overflow pipe which leads to another, bigger collection vessel.

Weighing rain gauge

Some gauges record the weight of any precipitation that falls – the simplest ones use a container that moves a pen which makes marks on a rotating drum. More modern ones are electronic and use simple sensors to take measurements.

Tipping bucket rain gauge

The tipping bucket rain gauge measures the amount of rain that falls in a given period – usually this is set to one hour. It collects the rain in much the same manner as other gauges; however, it then ducts it into a small receiver. When a specified volume of water has been collected, the receiver tips over, and sends a signal to a recording device.

Wind chill

Wind chill is a term used to describe the way that cold air and wind combine to make it feel colder than it really is. The effect is due to the way that moisture evaporates away from the skin, cooling it down. For those who climb to high altitudes or venture near the poles, minimizing it is vital.

High-tech measuring methods

Some weather parameters cannot be recorded directly by basic instruments, and need much more advanced methods. Technology provides a solution in the form of high-technology equipment such as advanced radar and weather satellites. The earliest method used, however, was the weather balloon.

Weather balloons

One of the problems meteorologists face is that taking measurements from high in the atmosphere is very difficult. The idea of using weather balloons – also known as 'sondes' – to carry recording instruments to high altitudes was first tested in France in 1892. Back then they were vast structures that were several thousand cubic feet in volume. They carried many different instruments, including barometers which measured barometric pressure, thermometers for temperature, and hygrometers to record humidity. Since there were no means to transmit the recorded data back to the base stations, the weathermen had to wait until the balloon came back to Earth to retrieve their measurements. This could be a difficult task as the balloons often drifted several hundred miles from where they were originally launched. This was later solved by a German meteorologist who made balloons which burst once the instruments had taken their measurements – parachutes were then used to return everything safely back to the ground. In the 1930s, these were superseded by newer versions which used radio transmitters to send continuous information back to base. Called 'radiosondes', these revolutionized the process of taking high altitude atmospheric recordings. Since then, several improved versions of weather balloons have been developed – over the years these have permitted recording flights to be made at ever increasing altitudes. They are now able to reach elevations of up to around 27,400 metres (90,000ft) – they can also stay aloft for much longer periods. This means that they now travel much further than was possible previously, and that measurements can even be made far out over the oceans. These days satellites are used to communicate with the recording devices carried by weather balloons, and this technique provides an enormous amount of invaluable information to meteorologists all over the world.

Here a woman is launching a pilot balloon as part of the programme to record atmospheric conditions. This photograph dates from late in WWII when most men were fighting abroad. For many women this was the first time they had been given the opportunity to get involved with such work.

This radar installation was used by the National Severe Storms Laboratory during the late 1970s and 1980s to detect, analyse and map lightning strikes.

Radar

The introduction of weather balloons was a major step forward for meteorologists. However, technology rarely stands still, and it was not long after the advent of the radiosonde that more advances were made. These came about as the result of military developments – in the late 1930s it was clear that Germany was gearing up for warfare. The British knew that Hitler had amassed a huge air force, and that if the United Kingdom was going to be able to fight off an invasion, they would have to develop an early warning system. The engineers came up with a brilliant solution – they found it was possible to bounce radio waves off objects such as aircraft, and then detect the reflections. They developed this discovery into a system called 'RADAR' – short for 'Radio Detection and Ranging'. While this gave them the ability to determine the position and height of enemy warplanes, it was not long before further refinements also allowed them to make assessments of atmospheric conditions. Over the next few years dedicated weather radar networks were established all over the world. Radar is still being used to great effect to this very day. The United States, for example, has created the National Weather Service Next Generation Weather Radars, which is usually shortened to 'NEXRAD'. This is used both for day-to-day forecasting as well as for looking out for severe weather such as tornadoes and hurricanes. Another more specific system has been implemented by the Federal Aviation Administration. Known as 'Terminal Doppler Weather Radars', this is used at major airports to give early warning of possible localized threats to aviation such as microbursts.

Radar is good at assessing the make-up of clouds – pulses of radio energy which are sent into the atmosphere get reflected back by moisture-laden air. The more water droplets that are present, the stronger the returned signal. This information is then used to create detailed radar images, which are then used by meteorologists to work out where precipitation is likely to occur.

This Gulfstream IV is a high altitude, high speed, twin turbofan jet aircraft that was acquired by NOAA in 1996, to collect, process and transmit vertical atmospheric recordings in and around hurricanes.

Aircraft take atmospheric recordings

Ever since mankind first took to the air, meteorologists have used aircraft in order to take direct recordings of atmospheric conditions. At first, aircraft were only capable of reaching low altitudes, and the flight crews had to perform the recordings themselves. As a result, the recordings available to the weathermen were rather limited. As aircraft improved, however, larger crews could be carried, along with ever more sophisticated equipment. For many years, forecasting stations relied on the data they obtained. These days such methods have largely been superseded by satellites, but aircraft are still used for specific purposes, such as hurricane analysis.

Weather satellites

Scientists began work on rocket-launched satellites in the 1950s as a by-product of the missile technologies developed by Germany during the Second World War. The first man-made satellite was successfully sent into space on 4 October 1957, when the Soviet Union launched Sputnik 1. Launched from the Baikonur cosmodrome in Kazakhstan, it orbited the Earth every 98 minutes, and transmitted radio signals back to earth. Although this was a very rudimentary piece of equipment

The flight meteorologist is a key component in the airborne atmospheric recordings program.

This image shows an early TIROS satellite from about 1960. Later versions featured cameras mounted on their sides.

compared to the much more refined versions which came soon after, it kicked off the great space race which continued until the collapse of the Soviet Union. The United States responded by establishing National Aeronautics and Space Administration (NASA), and then by successfully launching Explorer I on 31 January 1958. This was a small satellite that took scientific instruments into outer space – these were used to study radiation. In 1960, the first of several Television Infrared Observation Satellites (TIROS) were launched. TIROS 1, as it was known, was used to assess the effectiveness of studying the Earth's weather systems from space. It soon proved that forecasts based on satellite-derived observations were a marked improvement over those that relied on traditional methods. This stimulated the American government to fund the development of a whole series of newer, improved satellite systems; today, the technology is very advanced indeed. These days, two quite different kinds are used for weather observations. The first are known as geostationary satellites. These are placed at a height of around 35,680 kilometres (22,300 miles) above the ground, where they travel at the same speed as the Earth turns. This makes them stay above the same spot all the time, in what is referred to as a 'geosynchronous orbit'. The National Oceanic and Atmospheric Administration (NOAA) operates a pair of these high-technology instruments which they refer to as 'Geostationary Operational Environmental Satellites', or 'GOES' for short.

OPPOSITE: **This is an artist's impression from 1965 of how a fully integrated environmental monitoring system works. It includes satellites, balloons, ships, aircraft and buoys as well as data reception and processing facilities.**

The first GOES satellite was launched on 16 October 1975, and it soon began supplying

weathermen with invaluable information. This lone satellite was joined by another in 1977, but both have since been replaced by others. Currently, there are two GOES satellites in use – these are GOES-10 (also known as 'GOES-West') and GOES-12 (also known as 'GOES-East'). Although GOES-11 is orbiting the Earth, it is being kept out of operation as a back-up in case either of the others fails. The GOES satellites carry two main instruments – these are an imaging radiometer and an atmospheric sounder. The imaging radiometer focuses on specific parts of the Earth where it senses any radiant and solar reflected energy. The atmospheric sounder measures certain important parameters such as temperature and moisture profiles, as well as cloud

The meteorological satellite designated ESSA 9 was launched at night by a rocket on 26 February, 1969.

temperatures and ozone distribution. Other on-board instruments include devices which help to provide data about solar activity. The other type of weather satellite is called a 'Polar orbiting satellite' – these are placed in much lower orbits, where they continuously circle the Earth from pole to pole. Since they are closer to the ground, they give a more detailed picture than the much higher geostationary satellites. They work in conjunction with the GOES satellites to provide very detailed atmospheric recordings to the National Environmental Satellite and Information Service (NESDIS), who then relay the data to a wide variety of weather-related establishments.

Improving hurricane forecasts

When a hurricane threatens to strike, most people do their best to get out of its way. Others, however, do exactly the opposite, and climb into special aircraft and head straight for its centre. This is because the information we receive from weather satellites about what is going on in the middle of a hurricane is incomplete. As a result, what may appear at first sight to be a relatively harmless storm, can actually turn out to be a much more dangerous hurricane. It is therefore vital to collect accurate atmospheric data from within the storm itself. This not only allows valid predictions to be made, but it also provides invaluable data for future research work. While the idea of deliberately flying into the centre of a hurricane may seem bad enough, the aircraft then continue right across the storm and out of the other side. They then turn around and go through the whole process again several times. The first hurricane flights began in 1944, when a major storm hit the United States along the coasts of Long Island and New England. Although around 50 people died as a result of it, it is thought that the death toll would have been nearer 600 without the advance warnings given by the flight crews. There are various groups of daring characters who risk their lives on these dangerous missions,

must know

Computers
To improve severe and extreme weather forecasts, the National Oceanic & Atmospheric Administration has recently spent $180 million on a range of supercomputers. These can perform 1.3 trillion calculations per second – a vast improvement from the 450 billion the researchers could previously achieve.

This image shows a plot of data from NASA's Quick Scatterometer (Quikscat). It shows the direction and intensity of surface winds across the Atlantic Ocean. Blue areas indicate the slowest winds, while orange shows the fastest.

including those from the Air Force Reserve's 53rd Weather Reconnaissance Squadron, as well as the National Oceanic and Atmospheric Administration. Since 1944, four research aircraft have crashed in the course of their investigations, and 36 crew members have lost their lives – all in the interests of giving the public better weather forecasts.

Forecasting fog

Although fog does not pose a threat to entire communities in the way that severe storms can do, it is nevertheless something that needs due consideration, especially by those who are at sea, in the air, or on the roads. Since travelling in fog can be very dangerous, making accurate forecasts is very important; however, to actually do so can be very difficult. Fog forms when there is a combination of still air and high humidity levels near the ground. There are a lot of possible factors that can lead to this situation, and meteorologists have to take all of them into account before making predictions.

Effect of cyclones and anticyclones

When forecasters are compiling their predictions, one of the first things they do is plot their atmospheric data on charts and look for major features. When a high pressure centre known as an anticyclone is present, they can be reasonably sure that there will be fair skies and little or no precipitation in its immediate area. In the northern hemisphere, the winds around an anticyclone blow in a clockwise direction. This means that where they come from the south they carry warm air, and those from the north are cold. The reverse is true in the southern hemisphere, where anticyclones rotate in an anticlockwise direction. If the weather maps show that low pressure centres called cyclones are present, then the winds will be blowing in the opposite direction to that for anticyclones. That is to say that in the northern hemisphere, they blow around an cyclone in an anticlockwise direction, and in the southern hemisphere they rotate clockwise.

Forecasting precipitation

Telling the public whether it is going to rain or snow before it actually happens is one of the key activities of weather forecasters, so getting the predictions right is very important. One of the key factors in determining whether precipitation is likely to occur is the presence of sufficient moisture at low levels. One of the best ways to assess this is to examine a map of the dew point temperatures of a given region. If these are close to their respective air temperatures, then it means that the moisture levels are high, and precipitation is possible. Before it can take place, however, there must also be some method of lifting the air so that condensation can happen. If there is a suitable front in the area, for instance, then the probability that precipitation will take place is high.

I want to know more?

Take it to the next level...

▶ **Early forecasting** 8–11
▶ **Birth of modern weather forecasting** 12–15

Other sources...
▶ **Make your own weather station by constructing some simple measuring instruments (see www. miamisci.org/hurricane/ weatherstation.html for more information).**

▶ **Watch the night sky to see if you can identify orbital satellites that might be carrying weather instruments.**

▶ **Look on the internet to see if you can find a web-cam link so that you can watch a satellite being launched into space by a rocket.**

weblinks...
▶ **www.cpc.ncep.noaa. gov/**

▶ **www.weather.gov/**

▶ **www.metoffice.gov. uk/weather/satellite/**

▶ **www.ghcc.msfc.nasa. gov/GOES/**

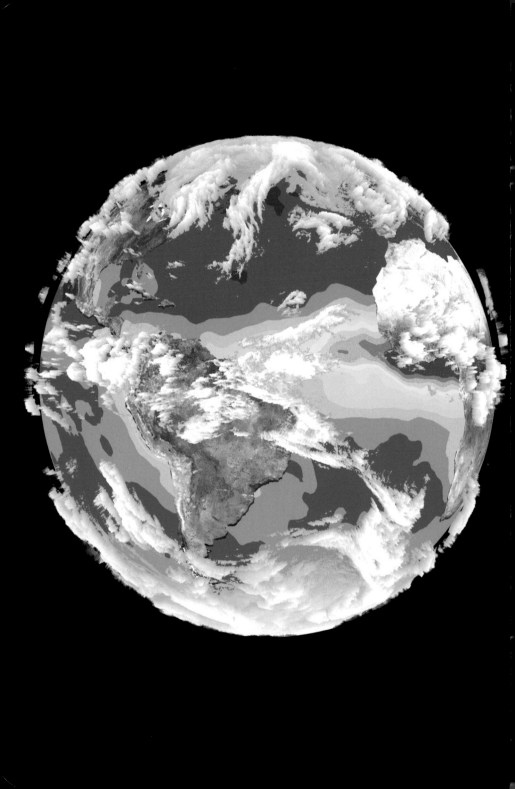

8 Mankind and the weather

The big question that mankind is currently facing up to is whether recent increases in global temperatures are due to human activity or not. If they are, we have to identify specifically what it is that we are doing that is causing these changes, and whether there is anything that we can do about it. This is such a politically charged issue, that before we look at the possible causes, it is worth examining the physical evidence as to what changes in our weather are actually taking place.

Global warming

Scientists generally agree that the Earth is warming up – in the last one hundred years, its surface temperature has risen by about 1–2°F. The rate of warming has been accelerating: the ten warmest years between 1900 and 2000 all occurred between 1985 and 2000.

A natural phenomenon or caused by man?

These increases in temperature have melted many snowfields, glaciers and vast areas of pack ice – this has, in turn, caused sea levels to rise between 10–20cm (4–8in) over the last century. Across the world there has also been an increase in precipitation by around 1 per cent; there have also been more extreme rainfall events in many parts of the world, including the United States. Additionally, atmospheric concentrations of many greenhouse gases have also increased since the industrial revolution began. The amount of methane, for instance, has more than doubled, carbon dioxide has gone up by around 30 per cent, and nitric oxides have increased by about 15 per cent. While these facts are not in question, there is furious debate as to what the causes are. It must be remembered that it is very hard to discuss the

The Seattle skyline at night. One of the hottest topics at the moment is just how much mankind's activities are altering the world's climate. Environmentalists say that our consumerist society is producing far too many atmospheric pollutants.

In order to work out what is going on with the global climate, it is necessary to study atmospheric conditions in every part of the world. Here, the NOAA Ship *Discoverer* ploughs its way through the Bering Sea.

situation whilst still remaining truly impartial, as must be done if one has an interest in getting at the truth. On the one hand there is the camp which claims that global changes are all down to human activity, and in particular that they are largely the result of excessive use of fossil fuels. On the other hand, there are those who say that this is unproven, and point to other possible causal factors, such as increased solar activity.

The facts

The picture is made far more complicated by the fact that there are many natural cycles to take into account – changes in global temperature have been occurring since before life began. Indeed, the myriad plants and animals on Earth have a significant influence on the atmosphere. Processes such as vegetative transpiration and organic decomposition release more than ten times the amount of carbon dioxide produced by mankind. Around 80 per cent of the atmosphere's methane is released by organic decomposition, and comes from rice paddies and swamps, as well as the digestive tracts of grazing animals and tropical termites. Volcanic eruptions also contribute enormous quantities of atmospheric pollutants, and have in the past been responsible for long periods of cold weather. Scientists have calculated that of the gases released by human activity in the United States, the combustion of fossil fuels accounts for about 98 per cent of the carbon dioxide, 24 per cent of the methane, and 18 per cent of the nitric oxides. The rest is the result of agriculture, industry, mining and so on.

The future

Whilst it is exceedingly difficult to determine exactly what is going on now, it is even harder to know what will happen over the next century or two – any calculations could be rendered invalid by just one large volcanic eruption. It is also impossible to know what is going to happen to the human population of the world; few scientists can agree on this one factor alone. Current projections which are based on what has happened over the last few decades suggest that global temperatures may rise by an average of 0.6–2.5°C (1–4.5°F) over the next fifty years, and by a further 1.4–5.8°C (2.2–10°F) in the next one hundred years. If these estimates are accurate, it is certain that global weather patterns will change significantly, and that in some regions life will become very difficult. Sea levels along the coast of the United States would rise by about 60cm (24in), and many low-lying areas would become permanently flooded. Some of the smaller Pacific Islands would disappear completely.

El Niño

One of the possible factors that might be involved in global warming is what is known as 'El Niño'. This is the name given to a short-lived warm oceanic current that develops every year along the coast of Ecuador and Peru around Christmas time. Usually, it only lasts a few weeks, however, every few years, it lasts a lot longer – from several months to a year at a time. The intervals between these prolonged events varies significantly – sometimes they may only be two or three years apart, whereas other times there may be up to seven years between them. When an El Niño current persists, the warm water it carries displaces nutrient-rich seas, and the local fishing industries more or less collapse. While this is bad enough, there are also consequences for the atmosphere over a far wider area. Tropical rains that normally fall over Indonesia get shifted eastwards, jet streams diverted, and monsoon patterns altered. The effects on

Over the years, hurricanes have wreaked enormous damage on human settlements. This is a US beach-front community prior to Hurricane Carol.

The same beach-front community *after* Hurricane Carol.

global weather systems have been known to cause droughts across the southern hemisphere, from Africa to India, Australasia, and South America. Severe rainfall has resulted in flooding in South and Central America as well as across the Caribbean and the United States. There have also been hurricanes in Tahiti and Hawaii. Not all the effects of El Niño are so marked, however. In the upper Midwest states and Canada, winter temperatures tend to be a bit higher than usual, and in central and southern California, northwest Mexico and the south-eastern United States, they are wetter than normal. In the summer, some regions of the United States are wetter than expected, whereas others are drier. Even though meteorological researchers have conducted extensive studies into the causes and effects of these irregular events, no clear conclusions have been drawn yet, and opinion remains divided.

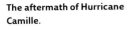

The aftermath of Hurricane Camille.

These are the remains of a large military aircraft after the passage of a tornado at Tinker Air Force Base, Oklahoma, USA on 25 March, 1948. This storm was responsible for the first ever broadcast of a tornado warning.

La Niña

Sometimes the warm waters resulting from El Niño events are replaced by cold water currents. When this happens, it is referred to as a 'La Niña' episode – these occur about half as frequently as El Niños, and often follow on directly behind them. Strong La Niña events happened in 1988–1989 and 1998–2001, and weak ones also occurred in 1995–1996.

El Niño Years

1902–1903	1905–1906	1911–1912
1914–1915		
1918–1919	1923–1924	1925–1926
1930–1931		
1932–1933	1939–1940	1941–1942
1951–1952		
1953–1954	1957–1958	1965–1966
1969–1970		
1972–1973	1976–1977	1982–1983
1986–1987		
1991–1992	1994–1995	1997–1998

The awesome power of a tornado is demonstrated by this 33rpm plastic record which has been blown into a telegraph pole.

Health and the weather

The weather can have all manner of effects on our health. In the more affluent areas of the world, bad weather generally causes few problems. However, in poorer places it can significantly weaken or even kill large numbers of people.

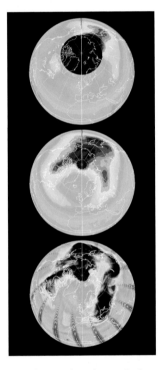

These images show changes in the ozone layer over the Arctic during the winter of 2004-05. The data was collected by the Ozone Monitoring Instrument (OMI) on NASA's Aura satellite. The top and middle images show the average total column ozone during January and March 2005; the lower image shows the same information for 11 March, 2005.

The symptoms

An extended period of rainfall, for instance, can cause widespread flooding. This can destroy drainage systems, resulting in the release of effluents into drinking water. When this happens, outbreaks of serious diseases like dysentery and cholera are commonplace. In the industrialized world, the weather's effects on health are usually much less serious – that is to say they are rarely life-threatening. Allergy sufferers, however, can suffer badly from maladies such as hay fever. As a result, things like the UV index, air quality and pollen counts are now major features of many weather forecasts.

The ozone layer and the UV index

The UV index is a measure of the amount of the sun's ultra-violet light that gets through the upper regions of the atmosphere to the ground. Mild exposure to ultra-violet light only causes sunburn, but prolonged periods can cause skin cancer and other serious medical problems. Usually, most of this harmful light is filtered out at high altitudes by a particularly reactive form of oxygen called ozone. Although this is a beneficial constituent of the upper atmosphere where it gathers in what is known as the 'ozone layer', if it accumulates near the ground it is toxic to life. It was discovered in the 1970s that the ozone layer was being depleted by atmospheric pollutants released through human activity. One of the main culprits was a class of chemicals known as chlorofluorocarbons (CFCs) – these are used as refrigerants in freezers and air conditioning units, and as propellants in aerosols. Where the ozone layer was thinning out, it was found that much higher levels of ultra-violet light were getting through, and so very strict regulations on the use of CFCs were imposed in the industrialized countries of the world. Although it appears that the ozone layer is recovering, research studies continue.

Where the ozone layer is thin – like here, in Australia – there is an increased danger of skin cancer – so it is vital to cover up.

The air quality index

The air quality index is a way of representing how clean or polluted the air is at any given time. In the United States, the

The air quality index

The six air quality index levels are:

'**Good**' To qualify for this rating the air quality index value must be between 0 and 50. It means that the air quality is satisfactory, and there is little or no air pollution.

'**Moderate**' To qualify for this rating the air quality index value must be between 51 and 100. It means that the air quality is acceptable, however, there may be a small health concern for people who are unusually sensitive – they may experience respiratory symptoms.

'**Unhealthy for Sensitive Groups**' To qualify for this rating the air quality index value must be between 101 and 150. It means that the air quality is good enough for healthy people, but may cause problems for those with either lung disease or heart disease.

'**Unhealthy**' To qualify for this rating the air quality index value must be between 151 and 200. It means that the air quality is poor, and even healthy people may suffer from health problems.

'**Very Unhealthy**' To qualify for this rating the air quality index value must be between 201 and 300. It means that the air quality is bad enough to cause serious health problems.

'**Hazardous**' To qualify for this rating the air quality index value must be over 300. It means that the air quality is so bad that emergency warnings for the entire population are likely.

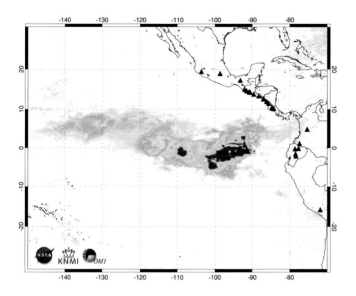

This image shows the distribution of sulphur dioxide which resulted from the eruption of the Sierra Negra, a 1,500m (4,890ft) high volcano on Albermarle Island, Galapagos.

index covers five major air pollutants – these are ground-level ozone, carbon monoxide, sulphur dioxide, nitrogen dioxide and airborne particulates. The index goes from 0 to 500, with a value of 50 designating good quality air, and 300 or above indicating that the air is hazardous to health. Air quality is an issue that we all need to care about. Adult humans breathe about 3,000 gallons of air a day, and any chemicals in it are easily absorbed into the body. Children and the elderly are particularly at risk, and those who are exposed to hazardous levels may get a burning feeling in their eyes, an irritated throat, or in extreme cases, experience breathing difficulties. In the long-term, more serious conditions such as cancer may arise, and even sudden death can occur.

Pollen counts

Pollen is produced by all flowering plants as part of the reproductive cycle – it is released into the air in the form of minute particles which are then carried elsewhere by winds.

OPPOSITE: **This image from the Moderate Resolution Imaging Spectroradiometer (MODIS) on NASA's Terra satellite shows atmospheric haze across the Eastern United States.**

Although most people are unaffected by it, countless millions experience a form of rhinitis – that is, a reaction that occurs in the eyes, nose and throat when airborne irritants or allergens trigger the release of a chemical called 'histamine'. This causes inflammation and fluid production in the fragile linings of nasal passages, sinuses and eyelids. Consequently, sufferers dread the onset of spring and summer as they fall victim to what is known as 'hay fever' or 'rose fever'. They can experience days, weeks, or even months of runny noses, eyes that continually water, and seemingly endless bouts of sneezing. Pollen counts are published by many weather forecasting services, and reflect

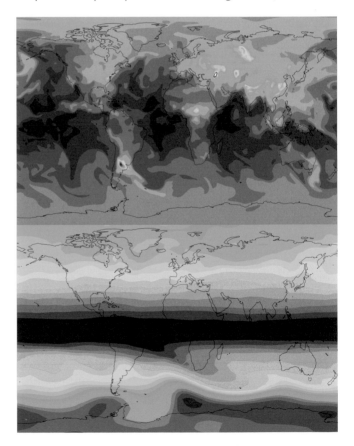

These two images show the distribution of methane across the globe – at the surface (upper) and in the stratosphere (lower). They were created from data compiled from a NASA computer model.

the current levels of airborne allergenic pollen – these are the microscopic agents that trigger hay fever attacks. This information can then be used to help sufferers to plan their schedules and, where necessary, seek medication. During the course of the year the risk varies tremendously depending on a person's particular susceptibility and geographic location. Some sufferers only react to pollen from one or two specific plant species, and may therefore only experience symptoms for short periods of time. In more extreme cases, others can be affected for much of the year.

Seasonal variations

Many trees start to produce huge quantities of minute pollen grains as early in the year as January. These include such species as hazel and alder. As winter gives way to spring, other species such as elm, ash and willow also begin to release their pollen. Fortunately for the majority of hay fever sufferers, these early species only produce adverse reactions in a small number of people. In Europe, it is not until the silver birch starts to flower in April that larger numbers of people begin to experience the dreaded symptoms. About 25 per cent of sufferers react badly for about four weeks, with the worst days being when the weather is warm and dry, with light winds.

After the birch trees it is the turn of oaks, but although many people start displaying the classic symptoms of hay fever at this time, it is usually grass pollen to which they are actually reacting. In North America, the main sources of allergenic tree pollen are oak, ash, elm, hickory, pecan, box elder and mountain cedar. The peak time for grass pollen is June and it may cause problems right through until the end of August. While grass pollen affects about 95 per cent of all hay fever sufferers, others are also affected by pollen from various wild flowers such as ragweed, which can produce pollen until late September.

must know

Migration
Many animals are migratory; that is, they move over large distances according to some kind of regular pattern. Usually, these movements are triggered by atmospheric conditions, but it has been observed that as the climate heats up some species are now able to survive through the winter in areas that they have not been able to before.

Mankind's influence on the weather

Over the last few years there has been a growing awareness that the global climate is changing. The polar ice caps are melting, glaciers are receding, and deserts expanding. It is therefore a matter of urgency that we work out just what is going on.

Deforestation

Although there are fierce arguments as to what is responsible for the observed increases in global temperatures, no-one can deny that deforestation has had a devastating effect on the environment. It has been estimated that less than half of the planet's surface remains unaffected by commercial or agricultural activities. Recent scientific studies have suggested that when tropical rainforests are destroyed, it can affect rainfall in faraway places. For instance, computer modelling has shown that deforestation in the Amazon Basin can influence the amount of rain that falls in Texas, while denuding the forests in Central Africa impacts on the level of precipitation experienced in the upper and lower US Midwest. These effects are not just limited to the United States. On the other side of the world, deforestation in Southeast Asia may change rainfall patterns on the Balkan Peninsula, which is several thousand miles away on the other side of the continent. In some cases the amount of precipitation that falls increases, while in others it decreases. The variations can also be more marked at certain times of year.

The results of the studies discussed here are not the end of the story, however; they merely highlight certain consequences of deforestation. In reality, large rainforest areas such as the Amazon are inextricably linked to global weather systems. Before mankind began cutting down the forests, they covered enormous areas, and the vegetation within them played a pivotal role in the hydrological cycle. When the trees are removed, the cycle becomes severely disrupted, and the once lush areas often become arid deserts. About two-thirds of the rain that falls across the world does so in the tropics. As it descends it releases a lot of energy in the form of heat. This has a direct impact on the way that the atmosphere behaves, right across the entire planet.

Deforestation has many other significant effects on the weather as well. Since trees absorb carbon dioxide as they grow, they lock in atmospheric carbon; as a result, they are what is known as 'carbon sinks'. That is, they remove one of the most

Deforestation is changing local weather systems.

Deforestation facts

► At the start of the 19th century there were 2.9 billion hectares of tropical forest worldwide; there are now only 1.5 billion hectares left.

► In the 30 years between 1960 and 1990, more than 445 million hectares of tropical forest were cut down. During this period, the Asian continent lost almost a third of its tropical forest.

► Nearly 90 per cent of West Africa's rainforests have been cleared.

► It is thought that over 90 different Amazonian tribes disappeared during the 20th century.

significant greenhouse gases from the atmosphere. While trees are growing, they also shade the ground, allowing it to retain moisture, as well as releasing large amounts of oxygen into the air. All these factors have a beneficial effect on the environment. On the other hand, cutting down trees exposes the ground to the sun's rays, reduces the amount of oxygen being released, and if the timber is burned, releases more carbon. Deforested areas are also often used for raising cattle, which themselves generate large amounts of methane – another major greenhouse gas.

Acid rain

Mankind's activities have had many deleterious effects on the environment over the years – acid rain being one of them. This is a generalized term which is used to describe the many ways that acidic compounds can be precipitated from the sky. These chemicals can be put into two groups – 'wet' and 'dry'. Wet forms include acidic rain, fog, and snow, whereas dry forms are held in the atmosphere as gases and particulates. Both pollute anything they come into contact with – living organisms may be poisoned or killed, and structures such as buildings or cars irreversibly damaged. The primary cause of acid rain is the use of fossil fuels – mostly in automobiles and coal-fired power stations. When oil or coal is burned, it releases a variety of pollutants, of which sulphur dioxide and nitric oxides are the main constituents. These then react with components in the atmosphere to form weak acids, including sulphuric and nitric acid. Over time, these toxic chemicals accumulate in waterways or as wind-blown dust, and as a result can be transported vast distances from the places where they were first released into the atmosphere. The problem began with the age of industrialization, and until recently it continued to get worse and worse. Lakes and rivers became seriously polluted, and fish species such as the brown trout more or less disappeared across entire regions. Large areas of forested land were also badly damaged, as were many sensitive habitats – particularly those at high altitudes where the soil had insufficient minerals to offset the extra acidity. It is not just the natural environment that suffers; where acid rains are prevalent – as indicated by the Air Quality Index – they also contribute significantly to poor human health. For the last decade or so governments around the world have been trying to reduce the amount of pollutants that are being released into the air; as part of this process, they have instituted several bodies to closely monitor the emissions levels. In some places that were once badly affected by acid rain there have been dramatic improvements, and it is heartening to see that many displaced species – including the brown trout and a number of delicate plants – have begun reappearing.

The deleterious effects of acid rain can clearly be seen in this photograph of severely denuded spruce trees.

Mankind's attempts to control the weather

Over the years, there have been many suggestions put forward as to how we might be able to control the weather. Some of these have bordered on the insane, others have been based on good scientific theory. Many have been tried out with no demonstrable results, but others are still being studied to determine how successful they have been. All meteorologists agree that it will never be possible to completely control the weather; however, there are those that still think that it might be possible to modify certain aspects of it. It has been argued that since hurricanes require warm water to form, if we could cool the water down, it would prevent these powerful storms from developing. One of the proposed ways of doing this is to tow icebergs into the relevant areas and then allow them to melt. Experts on the subject point out that the mass of water

During the 1960s there were many attempts to alter the way the weather behaved. Here, the crew of a Weather Bureau DC-6 are lined up proudly during Project Storm Fury – a hurricane cloud-seeding experiment.

which would need to be cooled is so vast that an iceberg would have no discernible effect on water temperatures over the required areas. Another unusual proposal concerns using satellite-based microwave generators to heat the air in storm systems in order to stop tornadoes forming – again, experts have ruled this concept out as being unworkable.

Probably the most well-known of mankind's attempts to control the weather is what is known as 'cloud-seeding'. This received a lot of media attention in the 1960s and 1970s, and involves using chemicals – usually silver iodide – to either increase or decrease the amount of rain, snow, fog or hail. The principle is that fine particles of the seed chemicals are released into clouds, either from aircraft or by rockets, to speed up condensation. This creates tiny ice particles that then fall as rain or snow. The official word on this is that there is 'no convincing scientific proof that cloud-seeding works'. Nevertheless, many American states continue to invest money in such schemes.

Renewable energies

With fossil fuels fast running out, it is vitally important that we find new ways of generating electricity to run our homes and industries. While nuclear power has a lot of things going for it, there are too many downsides for it to become a politically acceptable solution, at least in the short-term. There are abundant sources of energy around – the problem is that most of them are difficult, if not impossible, to harness with our current technologies. A good example of this is the solar radiation that creates the polar aurora; there is an incredible amount of energy there, but we do not know how to use it yet. It is likely that it will be many decades, if not centuries, before we identify a successful means of doing so. In the meantime, there are several more accessible sources of energy that are also inexhaustible or 'renewable'. Here we will examine four of them – these are wind, solar, tidal and hydroelectric power.

Wind power

Wind power is one of the more popular forms of renewable energy. It has been used for thousands of years, with the first use probably occurring on primitive sailing boats. Since these times many other methods of exploiting wind power have been developed. It is likely that the first windmills were built to pump water in Persia between 500 and 900 AD – later ones were also constructed for grinding corn. Since those early times, different methods of using wind power continued to be developed right up until the advent of steam power. Shortly after this electricity was discovered, and apart from sailing yachts and a few small-scale water pumps, wind power largely went out of fashion for about a hundred years. There were several experiments conducted into constructing commercial wind generators during the 20th century, but none of them led to wide-scale adoption of the technology. These included trials by Denmark, France, Germany, Great Britain, Russia and the United States. These days, wind power is better placed to become a viable replacement for fossil fuel-driven power, and although it does have its detractors, so does every other option. The two main factors in its popularity are that it is relatively cheap, and once the sites are operational, does not pollute the environment to any significant degree.

How is wind power harnessed?

The energy is extracted from this truly inexhaustible source by using the wind to turn large turbines – these are connected to generators which then produce electricity. When several of these structures are located close to each other, the site is known as a 'wind farm'. There are two basic forms of wind turbine – those that are located on land, known as 'onshore', and those that are sited at sea, which are known as 'off-shore'. In areas where there are consistent winds, this form of power generation is very attractive. Unfortunately, these winds are often a feature of our coastlines, which is just where many people do not want to see large arrays of turbines situated. Off-shore wind farms are more expensive to construct than those on land, but at least they are usually out of sight. In places such as the United Kingdom, it has been calculated that off-shore power alone could supply the region's needs three times over. The resultant electricity is also relatively cheap, costing much the same as that generated by conventional means. Many environmentalists are in favour of building wind farms as a way of reducing greenhouse gas emissions. One wind turbine alone can reduce emissions of CO_2 by over 4,000 tons a year, but in spite of this, there are still a lot of people who are against their use. Whilst these groups cite many different arguments to support their point of view, at the end of the day the main issue is that they do not like the look of the tall turbine towers.

Wind turbines were developed with the best of intentions. Unfortunately, they have attracted a lot of criticism – mainly from people who are unhappy because their view of the landscape has been changed.

Solar power

Solar power facts

▶ There is so much energy in the sunlight that reaches the Earth that one hour's worth could supply the entire planet for a year.

▶ The silicon content in one ton of sand could be used to make enough photovoltaic cells to produce as much electricity as burning 500,000 tons of coal.

▶ In the United States, over 10,000 homes are entirely powered by solar energy.

The term 'solar power' is used to describe any means of directly obtaining energy from the Sun. Since its energy is vital for life, any form of generation that uses plant or animal matter – such as oil, wood or coal, is effectively an indirect form of solar power. These pages, however, are concerned only with the direct methods. The Sun produces enormous amounts of energy – of that which reaches the Earth, about a fifth is absorbed by the atmosphere and just over a third is reflected back into space by clouds. In temperate regions, an average of somewhere between 125 and 375 watts per square metre actually reaches the ground. A certain amount of this can be captured and used to heat water or generate electricity. The most common method of doing this is to use what is known as a 'photovoltaic (or photoelectric) cell' – these are specialized semiconductors which turn sunlight into weak electric currents. When enough of these are assembled in a large array called a 'solar panel', useful amounts of electricity can be produced. Until the technology improves, however, they are too expensive to be used for large-scale purposes, although there are several experimental power plants that are connected to national grids in America and Europe.

Over the last fifteen years or so, photovoltaic cells have approximately halved in price, and so it may not be too long before they become economically justifiable for more generalized use. There have also been many refinements to the technology as the costs of energy derived from other sources continue to rise. Solar power is dependant on there being sufficient sunlight for the panels to function – this is therefore a method of power generation that is most suited to locations that receive a lot of sun. One method of increasing the amount of light is to construct a large array of mirrors so that sunlight from across a wide area can be focused onto the panels. Places like Southern California are ideally situated to exploit the technology – there is not only plenty of good weather, but also a lot of open

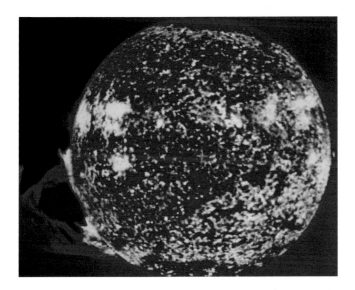

In this dramatic image of the sun a huge solar flare can be seen at the bottom left. Such events generate massive radiation bursts which have been responsible for power and communications failures on Earth.

space in which to construct the extensive solar farms. Since solar panels do not produce any greenhouse gases once they are in place, many governments are trying to encourage their use via tax concessions, grants, or other incentive schemes.

Tidal power

Another inexhaustible source of energy is the sea. The respective gravitational pulls of the sun and moon cause regular tides that offer the potential for cheap, clean power generation. This is known as 'tidal power', and it works in two main ways. The first method exploits the difference in the water's height between the high and low tide marks; however, this requires the construction of special sea dams or artificial lagoons. This has significant environmental downsides, and so is not the preferred solution. The other method is considered to be much more practical – it uses the strong currents that flow as the result of tidal movements. Identifying suitable sites is absolutely critical to their success, however, and as a result many potential sites are being surveyed across the world. One of the biggest advantages of tidal power is that oceanic currents are much more predictable than wind or solar power which rely on the right atmospheric conditions to work efficiently. Their operation is, however, restricted to the state of the tide, and as a result they can only generate electricity for between six and 12 hours a day. If this happens to coincide with the times of day when the demand for electricity is high, the power generated can be used directly. If not, it has to be stored in some manner, which is inefficient and costly. It also means that other forms of generation have to be available to supply the power needed when the tides are slack. Although tidal power generators are cheap to run, they are very expensive to build, and so it can take a long time to recover their construction costs.

Hydroelectric power

Hydro-power is that which is generated by extracting the potential energy from falling water – this is a practice that has existed since ancient times. The Romans, for example, used waterwheels extensively, and since then mills across the world have continued to be driven by water turbines. It was not long after the discovery of electricity that the first hydroelectric generator was built, and its success led to a proliferation of such plants. By the end of the twentieth century, between a fifth and a quarter of all the electricity generated in the world came from hydroelectric sources. Norway, for instance, produces virtually all of its electricity in this manner. One of the main advantages of hydro-power is that it can be supplied on demand – water is simply stored in elevated locations, and then released through

A hydroelectric power plant in Scandinavia, where this form of power is widely produced and used.

turbine-driven generators when needed. Another significant issue is that hydroelectric plants tend to need very little maintenance – this makes them cheap to run. Initial construction, however, can be extremely expensive, both economically and environmentally. This is because it is usually necessary to dam up entire valley systems to provide the required head of water. An enormous amount of civil engineering is therefore needed, and this is incredibly expensive. Much worse than that, however, is the fact that whole species can be pushed into extinction as the result of habitat loss. This occurs in three main ways: firstly, it is due to the vast areas of land that are flooded; secondly, because dams prevent creatures such as fish making their way to their breeding grounds; thirdly, because dams reduce the flow of water to such an extent that water levels in downstream areas can fall to the point where sensitive species may not be able to survive. There can be significant human repercussions, too – local people may be displaced, and ancestral, cultural or spiritual sites lost.

Although hydroelectric generators do not themselves release greenhouse gases, recent studies have shown that the reservoirs which often supply them can do so. This is especially true in tropical regions where rotting vegetation in and around these artificial lakes can be responsible for producing significant amounts of carbon dioxide and methane. This is because during the hotter parts of the year, the water levels fall, and large numbers of plants quickly grow in the damp soil; when the levels rise again, however, these are swamped and die. As they go through the decomposition process, marsh gas (methane) is released, along with carbon dioxide. Methane is particularly bad as it is more than twenty times worse for the environment than carbon dioxide.

want to know more?

Take it to the next level...

▶ **Changing climates** 15
▶ **Hurricanes** 94-9
▶ **Hurricane warnings** 100

Other sources...

▶ **To find out more about changing weather, buy a copy of** Currents of Change: El Niño's Impact on Climate and Society, **Michael H. Glantz.**

▶ **Find out where to access information on the air quality index in your area.**

weblinks...

▶ **www.epa.gov/global warming/kids/**

▶ **www.nas.nasa.gov/ About/Education/ Ozone/ozonelayer.html**

▶ **www.nasa.gov/ centers/goddard/news/ topstory/2005/deforest _rainfall.html**

▶ **pubs.usgs.gov/gip/ acidrain/2.html**

▶ **www.usbr.gov/power/**

Glossary

A

Acid rain This is the term used to describe rain that has been polluted by emissions of sulphur dioxide and nitrogen oxides from industrial activity or volcanic eruptions. It can kill or weaken plant and animal life.

Anemometer This is an instrument used to measure wind speed.

Atmosphere This is the term used to describe the mixture of gases that surround the Earth and are held in place by gravity. There are several different layers – these are the stratosphere, troposphere, mesosphere and thermosphere.

Aurora Borealis This phenomenon is also referred to as the 'Northern Lights'. It is an optical effect caused by interactions between solar radiation and atmospheric particles that occurs near the poles.

B

Barometer This is an instrument used to measure air pressure.

Beaufort Scale This is a scale used to measure wind speed.

Blizzard This is a severe snowstorm that lasts for four hours or more, and is often accompanied by strong winds and low temperatures. Such conditions can lead to significantly reduced visibility on the roads, the sea or in the air.

C

Climate This is the term used to describe the typical meteorological conditions of a place or region.

Cloud This is a visible assemblage of minute water and/or ice particles in the atmosphere.

Cumulus This is a type of cloud that is familiar over most of the world, often in the form of mounds or towers and having sharply defined edges.

D

Doppler Radar This is a form of radar that is used by meteorologists to measure the position and speed of atmospheric disturbances.

Drought This is an extended period of dry weather.

E

El Niño This is the term used to describe a regular warming of the equatorial waters in the Pacific Ocean that usually occur every three to seven years. They cause variations in regional weather patterns.

F

Fog This is the term used to describe a cloud that is at ground level. Under warm conditions, it is composed of minute water droplets; however, when it is very cold, these freeze into tiny ice crystals. Severe fog can be a major hazard due to reduced visibility.

Frost This is the term used to describe soft white ice crystals or frozen dew drops that settle on objects when the surface temperature falls below freezing point.

Fujita Scale This is a scale used to measure wind speeds based on the amount of damage they cause to structures such as buildings or trees.

G

Gale This is the term used to describe strong winds that have a speed of between 28 knots (32 mph or 51km/h) and 55 knots (63 mph or 102km/h).

Greenhouse Effect This is the term used to describe the way that the atmosphere traps the sun's heat. Without this effect, global temperatures would be much lower, and life would not be possible as we know it.

H

Hail This is the term used to describe a form of precipitation where small pieces of ice fall instead of rain or snow. In extreme circumstances these can be anywhere up to the size of grapefruit.

Humidity This is a measure of the amount of water vapour in the air.

Hurricane This is the term used to describe a tropical storm where the wind speeds are between 64 knots (73 mph or 117km/h) and 240 knots (150 mph or 414 km/h). In different parts of the world they are also referred to as typhoons or tropical cyclones. These storms can wreak an enormous amount of damage, and many tens of thousands of people have been killed by them over the years.

Hygrometer This is an instrument used to measure humidity.

I

Isobar This is the term used to describe a line connecting points of equal pressure.

Isotherm This is the term used to describe a line connecting points of equal temperature.

J

Jet Stream This is the term used to describe strong high-altitude winds that flow in narrow streams.

K

Kilopascal This is a unit of measurement used for atmospheric pressure.

L

Lightning This is the term used to describe the various types of visible electrical discharge that are produced by thunderstorms.

M

Meteorology This is the study of the atmosphere and atmospheric phenomena.

Mist This is the term used to describe a suspension of microscopic water droplets in the air. It is similar to fog, but less dense.

Mesosphere This is the layer of the atmosphere that extends from 50 km (30 miles) to 80 km (50 miles) above the ground.

N

Nocturnal This term is used to refer to an event or activity that takes place at night.

O

Ozone This is a gas that occurs naturally in a layer in the stratosphere where it absorbs harmful ultraviolet radiation. It also occurs at lower levels as the result of pollution, where it is toxic to life.

P

Precipitation this term is used to refer to water falling from the sky in the form of hail, mist, rain, sleet, and snow, but not dew, fog, or frost.

Glossary

R

Rainbow This is an optical effect caused when sunlight is refracted and then reflected by raindrops; this breaks the light up into its component colours.

Rain Gauge This is an instrument used to measure the amount of rainfall in a given location.

S

Stratosphere This is the layer of the atmosphere that extends from 10 km (6 miles) to 50 km (30 miles) above the ground.

T

Thermosphere This is the layer of the atmosphere that extends upwards from 80 km (50 miles) above the ground.

Thunderstorm This is the term used for storms where electrical discharges occur as lightning – these are accompanied by loud bangs known as thunderclaps.

Tornado This is the term used to describe a violent funnel-shaped wind vortex which forms from the base of a cumulonimbus cloud during severe thunderstorms.

Tropopause This is the upper boundary of the troposphere.

Troposphere This is the layer of the atmosphere that extends into the air from the ground up to 10 km (6 miles).

W

Waterspout This is the term used to describe a tornado that occurs over water.

Wind Chill This is the term used to describe the cooling effect caused by a combination of wind and low temperatures, and is measured by what is referred to as the wind chill factor.

Need to know more?

If you are keen to delve deeper into the mysterious world of weather, there is a wealth of information available, not only in books but also on numerous websites.

Bibliography

Introduction to Weather Pamela Bliss. National Geographic Books. 2004.
Scanning the Skies: A History of Tornado Forecasting Marlene Bradford. University of Oklahoma Press. 2001.
Extreme Weather: A Guide & Record Book C. C. Burt. W. W. Norton & Company. 2004.
Storms Andrew Collins. National Geographic Books. 2002.
The Tornado – Nature's Ultimate Windstorm T. P. Grazulis. University of Oklahoma Press, 2001.
Extreme Weather Glen Phelan. National Geographic Books. 2004.
Rainbows, Halos, and Glories Robert Greenler. Cambridge Univ. Press, Cambridge. 1980.
The Rough Guide to Weather Robert Henson. Rough Guides, 2002.
The Stories Clouds Tell Margaret A. LeMone. American Meteorological Society. 1993.
The Audubon Society Field Guide to North American Weather David M. Ludlum. Knopf, 1991.

Cambridge Guide to the Weather Ross Reynolds. Cambridge University Press, 2000.
Hurricane Watch: Forecasting the Deadliest Storms on Earth Dr. Bob Sheets & Jack Williams. New York: Vintage. 2001.
The USA TODAY Weather Book Jack Williams. Vintage Books. ISBN 0679776656. 1997.

Websites

http://nws.noaa.gov/
This site is for the National Weather Service, part of the National Oceanic and Atmospheric Administration.

www.ncep.noaa.gov
This site is for the National Centres for Environmental Prediction, which is also part of the National Oceanic and Atmospheric Administration.

http://www.weather.com/index.html
This website provides information on local as well as global weather, and has

many informative articles, features and links.

http://airnow.gov/
This site was established by a number of official agencies to provide the public with easy access to information about air quality across the United States.

http://www.noaa.gov/tornadoes.html
This is another website established by the National Oceanic and Atmospheric Administration – it provides a substantial amount of information about tornadoes.

http://www.tornadoproject.com/
This is a commercial site that gathers, compiles and makes tornado information available to weather enthusiasts, the meteorological community and emergency management officials.

http://www.nationalgeographic.com/siteindex/weather.html
This site provides an index to the National Geographic Magazine's articles on Weather and Natural Forces.

http://www.ucar.edu/
This website is for the National Centre for Atmospheric Research (NCAR) and the University Corporation for Atmospheric Research (UCAR) Office of Programs.

Weather Organisations & Societies

The American Meteorological Society
45 Beacon Street, Boston, MA 02108-3693 USA
Telephone: 617-227-2425
Fax: 617-742-8718
E-mail: webadmin@ametsoc.org
Website: http://www.ametsoc.org

The National Climatic Data Centre
Federal Building, 151 Patton Avenue, Asheville, N.C. 28801-5001 USA
Telephone: 828-271-4800
Fax: 828-271-4876
E-mail: NCDC.info@noaa.gov
Website: http://www.ncdc.noaa.gov

The National Weather Association
1697 Capri Way, Charlottesville, VA 22911-3534 USA
Telephone/Fax: 434-296-9966
E-mail: NatWeaAsoc@aol.com
Website: http://www.nwas.org/

The Royal Meteorological Society
Royal Meteorological Society, 104 Oxford Road, Reading, RG1 7LL United Kingdom
Telephone: +44 (0)118 956 8500
Fax: +44 (0)118 956 8571
Email: execdir@rmets.org
Website:
http://www.rmets.org/index.php

Index

Need to know more?

Acknowledgements

The majority of photographs and images used in this book are courtesy of the National Oceanic and Atmospheric Administration (NOAA) Photo Library. The homepage can be found at: www.photolib.noaa.gov

The remaining images were kindly supplied by the US Geological Survey (USGS) – www.usgs.gov; NASA; Ruth Stringer; and the author, Patrick Hook.

☾ Collins need to know?

Look out for these recent titles in Collins' practical and accessible need to know? series.

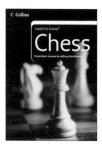

Other titles in the series:

Antique Marks
Birdwatching
Body Language
Buying Property in France
Buying Property in Spain
Card Games
Children's Parties
Codes & Ciphers
Decorating
Digital Photography
DIY

Dog Training
Drawing & Sketching
Dreams
Golf
Guitar
How to Lose Weight
Kama Sutra
Kings and Queens
Knots
Low GI/GL Diet
Pilates

Poker
Pregnancy
Property
Speak French
Speak Italian
Speak Spanish
Stargazing
Watercolour
Weddings
Wine
Woodworking

The World
Yoga
Zodiac Types

**To order any of these titles, please telephone 0870 787 1732 quoting reference 263H.
For further information about all Collins books, visit our website:
www.collins.co.uk**